T0191912

Springer Optimization and Its Applications

VOLUME 125

Aims and Scope
Optimization has been expanding in all directions at an astonishing rate during the last few decades. New algorithmic and theoretical techniques have been developed, the diffusion into other disciplines has proceeded at a rapid pace, and our knowledge of all aspects of the field has grown even more profound. At the same time, one of the most striking trends in optimization is the constantly increasing emphasis on the interdisciplinary nature of the field. Optimization has been a basic tool in all areas of applied mathematics, engineering, medicine, economics and other sciences.

The series *Springer Optimization and Its Applications* publishes undergraduate and graduate textbooks, monographs and state-of-the-art expository works that focus on algorithms for solving optimization problems and also study applications involving such problems. Some of the topics covered include nonlinear optimization (convex and nonconvex), network flow problems, stochastic optimization, optimal control, discrete optimization, multi-objective programming, description of software packages, approximation techniques and heuristic approaches.

More information about this series at http://www.springer.com/series/7393

Stamatina Th. Rassia • Panos M. Pardalos
Editors

Smart City Networks

Through the Internet of Things

Editors
Stamatina Th. Rassia
INSEAD
Fontainebleau, France

Panos M. Pardalos
Industrial and Systems Engineering
University of Florida
Gainesville, FL, USA

ISSN 1931-6828 ISSN 1931-6836 (electronic)
Springer Optimization and Its Applications
ISBN 978-3-319-87049-6 ISBN 978-3-319-61313-0 (eBook)
DOI 10.1007/978-3-319-61313-0

Mathematics Subject Classification (2010): 05C82, 05C85, 11Bxx, 11B05, 34B45

Softcover reprint of the hardcover 1st edition 2017

Printed on acid-free paper

This Springer imprint is published by Springer Nature
The registered company is Springer International Publishing AG
The registered company address is: Gewerbestrasse 11, 6330 Cham, Switzerland

Preface

Smart City Networks: Through the Internet of Things is composed of research results, analyses, and ideas, which focus on a diversity of interconnected factors relating with urbanization, its "smartness," and overarching "internet of things (IoT)." The latter refers to interconnected objects and devices – through computational operations – which can receive signals and actuate systems.

City functions and their respective network "smartness" correspond to the overarching paradigm of the IoT. However, city networks are highly complex. Therefore, their network organization, quality control, operations, and efficiency are open for further research. More specifically, due to the vast urbanization and the need for a faster pace of living – by means of taking shortest paths, minimizing walking and commuting distances – special factors that support the syntax of cities' smartness are yet to be identified. Although theories and concepts have been developed so far on the topic of IoT, applications in actual cities and their networks are open for further research. Hence, the question raised in this book is to identify a number of interconnected elements of the cities, which not only have an impact on its functionality but also importantly affect its livability.

Cities are both "living organisms" and "machines for living in," and therefore, there is a multitude of themes composing the groundwork upon which new ideas for future cities could be considered. Thus, in this book cities are examined as environments of systemic change, as unified technological advancements, computational models, natural environments, visuospatial platforms, and more. They are also analyzed according to their sustainability, properties, as well as navigation protocols.

Each chapter of this book studies a topic of interest in terms of promoting smart future city networks and thus presents the variety of interconnected elements related to the "Internet of things." The special characteristic of this book is that it acts as an open platform of expression for all its authors to present their views, understanding, and results on this complex yet fascinating topic. As the research on IoT is rapidly growing worldwide, in many different research directions and disciplines, the present book offers a framework where international experts present eminent work and highlight special approaches to this topic. In acknowledgment of all chapter contributions in this book, we would like to express our special thanks to all authors who participated in this collective effort.

Last but not least, we wish to acknowledge the superb assistance that the staff of Springer has provided during the preparation of this book.

INSEAD, France Stamatina Th. Rassia
University of Florida, USA Panos M. Pardalos

Contents

Contributors

Jihad Awad Department of Architecture, Ajman University, Ajman, UAE

Nick V. Baker Architectural Association, London, UK

Jo Ivey Boufford The New York Academy of Medicine, New York, NY, USA

Renata Paola Dameri Department of Economics and Business Studies, University of Genova, Genova, Italy

D. Fisher-Gewirtzman Faculty of Architecture & Town Planning, Technion-IIT, Haifa, Israel

Hermann Haken Institute for Theoretical Physics, Center of Synergetics, Stuttgart University, Stuttgart, Germany

Afaq Hyder Department of Architecture, Ajman University, Ajman, UAE

Adi Irfan Department of Architecture, National University of Malaysia, Bangi, Malaysia

Rupesh S. Iyengar Services Consultants, Bengaluru, India

C. Cartalis Physics Department, National and Kapodistrian University of Athens, Athens, Greece

Yan Li Center for Health Innovation, The New York Academy of Medicine, New York, NY, USA

Department of Population Health Science and Policy, Icahn School of Medicine at Mount Sinai, New York, NY, USA

A. Natapov Centre for Advanced Spatial Analysis, University College London, London, UK

José A. Pagán Center for Health Innovation, The New York Academy of Medicine, New York, NY, USA

Department of Public Health Policy and Management, College of Global Public Health, New York University, New York, NY, USA

Leonard Davis Institute of Health Economics, University of Pennsylvania, Philadelphia, PA, USA

Juval Portugali ESLab (Environmental Simulation Lab), Department of Geography and the Human Environment, Tel Aviv University, Tel Aviv, Israel

Francesca Ricciardi Department of Business Administration, University of Verona, Verona, Italy

M. Santamouris School of Built Environment, University of New South Wales, Sydney, NSW, Australia

Physics Department, University of Athens, Athens, Greece

Gaurav Sarin Delhi School of Business, New Delhi, India

Shahrokh Shamalinia Independent Researcher, Tehran, Iran

Larissa R. Suzuki Department of Computer Science, University College London, London, UK

Digital City Exchange, Imperial College Business School, London, UK

Dimitri Volchenkov Mathematics & Statistics, Texas Tech University, Lubbock, TX, USA

Artificial Intelligence Key Laboratory, Sichuan University of Science and Engineering, Sichuan, China

Designing Access to Nature for Residential Buildings

Nick V. Baker

Abstract The popular aphorism that 'nature is good for you' is explored by reviewing a number of studies measuring the impact of nature, and its deprivation, on subjects. These range from well-being in dementia patients to the development of cognitive and motor skills in preschool children. With sufficient evidence that access to nature is indeed good for you, and providing a pragmatic (if not rigorously scientific) definition of nature, the paper moves on to identify the key design parameters that have impact on our access to nature.

The work proposes a spatial model that consists of *zones*, *links and qualities*. *Zones* can be *inside*, *edge*, *near* or *far*, these corresponding to the building interior, the building envelope, the immediate surroundings and the distant landscape. Between these zones are *links* that are either *access or sensory*. All the above zones and links can be ascribed *qualities*. Whilst there is too little data at present to propose a quantitative calibration, the model may be useful to a designer for ordering and balancing various conflicting design decisions. Finally, other issues relating to nature are discussed. These include attracting wildlife into the near zone and facilitating gardening and pet-keeping.

1 The Case for Nature Access and Evidence from the Literature

People in the developed world spend 95% of their time indoors, whereas ten generations ago, our ancestors would spend most of their waking hours outdoors. And whilst we may have made the cultural adjustments to life indoors, genetically we have not, since the genetic changes in ten generations are insignificant. Our genes

N.V. Baker (✉)
Architectural Association, London, UK
e-mail: nickvbaker@aol.com

S.T. Rassia, P.M. Pardalos (eds.), *Smart City Networks*, Springer Optimization and Its Applications 125, DOI 10.1007/978-3-319-61313-0_1

evolved for survival in the wild plains and forests, where hazards from the climate, the landscape and the flora and fauna faced us every day. Could this explain our almost universal fascination with nature?[1]

Searches under 'access to nature' yield a huge body of work from the recovery of post-operative patients to the literacy of schoolchildren. The studies have often sought to identify corrective measures in areas of deprivation, and often in subjects which are either already vulnerable or showing signs of stress. For example, we find studies on the development of preschool children in poor urban environments and the value of gardening to dementia patients.

1.1 Proximity of Green Spaces

Many of the studies on the impact of nature on well-being are concerned with the proximity and physical accessibility of open green spaces, of varying levels of 'wildness' or more specifically biodiversity. In a comprehensive review by Strife and Downey (2009) of over 150 articles, which specifically reviewed studies on the impact of deprivation on youth development, access to nature was claimed to have physical, mental, emotional and cognitive benefits – and a positive effect on children's overall development.

To quote specific studies, Fjørtoft (2004) found that children using a forest as a play setting performed better in motor skills tests than children who used standard playground settings, although the latter contained modern play equipment. Grahn et al. (1997) found that children attending day-care facilities surrounded by natural woodlands had greater attention capacity and motor coordination than those attending centres surrounded by tall buildings. In a longitudinal study Wells (2000) showed that cognitive functioning improved when children moved into housing with nearby green spaces.

Other evidence that emerge from studies is that view of natural landscape can be beneficial in itself. Faber Taylor et al. (2002) reported a reduction in ADHD symptoms and improvement in academic performance amongst Afro-American schoolchildren, when the natural areas were used as part of the school curriculum. It was also claimed that green space generates social interaction and reduces violence, the continuing presence of 'ambient' nature was more effective than one-off experiences, and people were sensitive to the quality of natural environments and the safety of the users.

[1] Wilson's (1984) *biophilia hypothesis is based on the premise that our attachment to and interest in animals stems from the strong possibility that human survival was partly dependent on signals from animals in the environment indicating safety or threat. The biophilia hypothesis suggests that now, if we see animals at rest or in a peaceful state, this may signal to us safety, security and feelings of well-being which in turn may trigger a state where personal change and healing are possible* (https://en.wikipedia.org/wiki/Animal-assisted_therapy#cite_note-4).

Rather less research has been done on issues closer to the building itself. However Chalfont (2008), in a book 'Design for Nature in Dementia Care', identifies *building edge zones*, intermediate areas between inside and outside where the critical connection to nature can be made. He also reminds us of the phrase 'out of sight-out of mind' implying the need to provide a visual connection between the interior and the beneficial natural surroundings. We use a similar approach to Chalfont as a model for access later in the paper.

1.2 Views of Nature from the Building

As early as 1967, Markus (1967) reports the preference of people for distant views and for views that give them information about the weather. Ulrich (1984) reports the improved recovery from surgery in patients with distant views of trees compared with patients looking at a blank wall.

As part of her PhD thesis on Daylight and View, Helinga (2013) reviews over 50 papers under the heading of View Quality. A consistent assertion is that natural views are more highly valued. Of office workers, Kaplan (1993), (2001) reports that 'those with a view of nature felt less frustrated and more patient, expressed greater enthusiasm, and reported higher life satisfaction as well as overall health'. More specifically, Heschong Mahone (2003) reports a 10–25% better performance in mental tests amongst those with outside views and fewer complaints of fatigue and environmental discomfort.

Helinga goes on to incorporate the evidence in her own study by defining a scale of view quality from occupied offices which ranged from zero for no nature to 4 for 100% 'natural' view. However, qualifiers were added that enhanced the value of non-natural scenes due to factors such a depth and diversity. What is perhaps a more important conclusion of the study was that there was strong correlation between the perceived value and quality of daylight and the quality of the view. This is consistent with observation of higher satisfaction with other environmental parameters, such as thermal comfort reported by Heschong Mahone and with the claim by Nicol and Humphries (1973) that thermal satisfaction is influenced by views that have a climatic significance. For example, an occupant looking out on a snowy scene will be more tolerant of non-neutral (cool) temperatures and more prepared to take adaptive action to mitigate the discomfort.

To sum up, the literature provides overwhelming evidence to confirm that both physical and visual access to nature are an essential part of well-being.

1.3 A Working Definition of 'Nature'

Before considering a design response to the need for access to nature, we must clarify, for this purpose, what we mean by nature. Dictionary definitions tend to agree on 'all of the physical world that is not manmade'. Immediately we run into problems with this since most of the landscapes that surround our conurbations and settlements are to a degree man-made. And within those landscapes, many of the species – the crops and the animals – are themselves highly influenced by the breeding and selection of man. So too are the trees and plants found in our parks and gardens, as are the flowers which we place on our living room table.

Thus we adopt here a pragmatic definition that nature means that quality that gives an impression of being to a greater or lesser extent 'natural'. This could be at the extreme true wilderness and, at the other end of the spectrum, an urban landscape combining a minimal amount of vegetation and urban wildlife, but with distant views and sky, and a lack of obviously negative properties such as pollution and dilapidation.

1.4 Proposed Design Parameters

Following Chalfont's (2008) approach, we propose here a model with a systematic vocabulary that although may not be used fully quantifiably can be used as a structured checklist to assist design.

The model consists of ***zones***, ***links*** and ***qualities***. The total environment is divided into four zones – INSIDE, EDGE, NEAR and FAR. The zones are *linked* in two ways: ***Sensory*** and ***Access*** (Figs. 1 and 2).

INSIDE is the building interior. It can have nature content itself, e.g. indoor planting, indoor pet-keeping facility or even surrogate nature – e.g. photographs, paintings or sculpture.

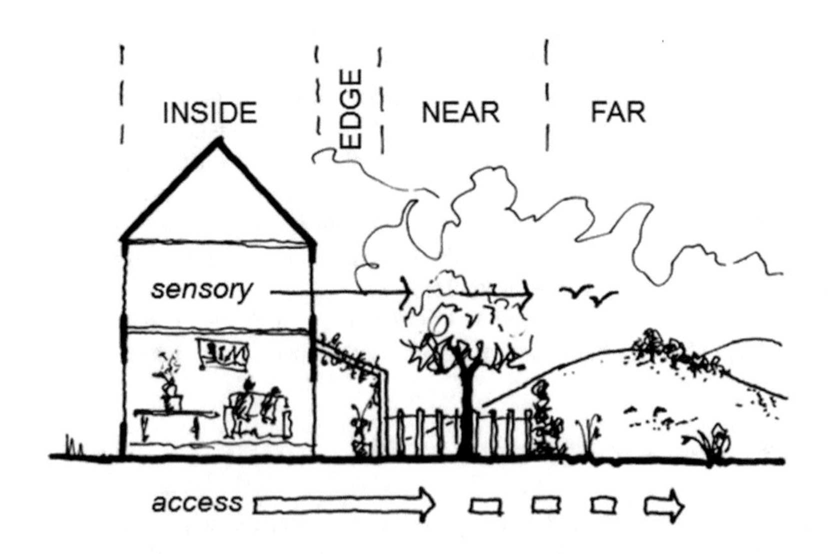

Fig. 1 Section illustrating the configuration of zones and links for low-rise residence with contiguous *NEAR* and *FAR* zone

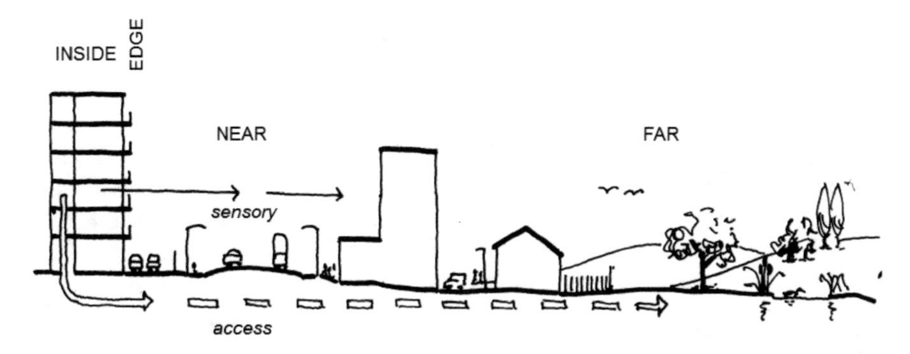

Fig. 2 Section illustrating the configuration of zones and links for a high-rise flat showing balcony as *EDGE*, the *NEAR* zone with no nature value and a remote *FAR* zone with high nature value. The ***sensory*** link to nature is compromised by the noise of the street and the obstruction of the buildings opposite. The ***access*** link is degraded by requiring a lift, crossing a busy road, and possibly an 'unsafe' neighbourhood

EDGE is the physical envelope and extensions of the building that links by means of doors, windows, balcony, porches and verandas. May have nature content itself.

NEAR is the space immediately outside the building such as a patio, yard or garden. It can have nature content and structures that promote nature – e.g. planters, ponds, bird table, etc. It will link with the INSIDE via EDGE. NEAR is usually private or semiprivate.

FAR is usually off-site, large in scale and often public, e.g. parks, farmland and wild land such as marshes, moors or mountains. May be *contiguous* with NEAR or *remote*.

Sensory is linked by vision, sound or smell. For sighted people, vision is the most important. Can be seen as a property of EDGE through which the link passes. The links can exist from INSIDE to FAR via NEAR or direct from INSIDE to FAR.

Access is the facility for the physical movement of the person from INSIDE to move to either NEAR or FAR. The link can exist between INSIDE and NEAR or INSIDE and FAR. If FAR is *contiguous*, then ***Access*** will be via EDGE and NEAR only. If FAR is remote, ***Access*** will be via EDGE and (possibly) NEAR and then an off-site area such as pathway, street, highway or even transport system, which will influence the quality of the ***Access***. Distance will also be a key parameter. Vertical distance is also critical for the INSIDE to CLOSE link. It is far less likely that a person will have good use of a garden if they are living in a fourth floor flat than a ground floor flat. It might be better to put resources into a shared patio garden at fourth floor level or even individual balconies.

QUALITY can be ascribed to all three environments and the links. In the case of the FAR environment, this will include scenic variety and biodiversity (the two being closely related). Clearly an area of rolling hills, outcrops of woodland and hedgerows is of greater restorative value both to view remotely and to walk amongst than modern ‘industrial’ farmland. The NEAR environment is more dependent on architectural elements and the way that the immediate surroundings to the building attract wildlife and horticultural activity. For the INSIDE environment in domestic spaces, the nature content is limited in architectural terms and relates more to occupant decisions about furnishing and contents. However in multi-residential blocks, there may be opportunities for indoor planting, water features and nature-related art, in spaces such as foyers, staircases and other communal areas. Many studies on indoor planting indicate that it has significantly beneficial effects.

QUALITY is just as important applied to links. ***Sensory*** INSIDE/NEAR and INSIDE/FAR mainly applies to windows. The importance of positioning windows in relation to views and to seating areas is well known to architects but is often neglected in response to demands for standardising layouts or even structural constraints. Special consideration should be given to supervision of children from within the house. A garden or patio that can be supervised from inside (possibly a kitchen) by a parent will be much more used by young children, that one where the parent has to enter the garden area in order to check up on the children.

QUALITY applied to the ***Access*** link is very important. It may be associated with the EDGE – i.e. the design and location of the door or doors into the *NEAR* environment, relative to the occupancy of the rooms. For example, good access would be where the main living area is linked to the garden or patio by doors also providing a visual link. This is not uncommon, often involving wide openings with sliding patio doors.

The QUALITY of the ***Access*** link to the FAR environment is equally important. If the FAR and NEAR environment is contiguous, the link will be short – i.e. a gateway in the boundary of the NEAR environment. This is highly desirable but probably a luxury that few urban dwellers enjoy. Far more frequently, the link will involve a journey through a built-up area of no nature value and generally used by the public. It may also involve dangers from road traffic and perceived or real dangers to children and vulnerable people from anti-social and criminal behaviour. Or it may simply be unpleasant due to ugly surroundings, noise and atmospheric pollution.

Another critical factor is the distance. In the Strategy for Improved Nature Access for London (Greater London Authority 2008), the target maximum distance is set at 1 km. This equates to about 10–15 min walking. Clearly, distance and environmental quality interact – a longer walk in pleasant urban surroundings could be regarded as being equivalent to a short unpleasant one. However, for vulnerable people, e.g. children, or the old, a single feature such as a dangerous road crossing, or a steep hill, might be critical.

2 Access for Wildlife

This refers to the access for wildlife *to* the NEAR zone. Sensory contact (usually visual) with active wildlife – birds, animals and insects – is greatly valued. Most birds have access by air, and even the common urban species such as pigeons provide a valuable nature resource in otherwise dire urban environments.

There is already much design guidance (English Nature 2005) available on provision for birds to nest and feed on or close to buildings. Indeed, much of the loss of common garden species is blamed on changes of building design which deny birds roosting and nesting places and changes in garden style that replace flowerbeds planted with flowering annuals and shrubs, with tarmac hard-standing for a car and paved patios with gas heaters.

Whilst designers cannot guarantee that the occupants will become avid gardeners, there is much that can be done to encourage it – the provision of planting opportunities, even in confined spaces such as balconies, and even providing surfaces with easy access where birds can be fed. Gardening activity will also have an impact on insect population, which in turn will attract birds. Whilst many insects may be of limited interest to the average occupant, and some regarded as pests, there is almost universal appreciation of butterflies, moths and dragonflies.

In public areas of high nature value, interpretation boards are found to have positive effect on nature appreciation. This principle could be extended to a more local level in multi-occupancy buildings, providing information on wildlife likely to be seen from the building.

3 Keeping Pets

Pet-keeping is widespread in Europe. In 2012 the European Pet Food Industry Federation claimed that about 25% of households have a dog or cat. Fish, reptiles, birds and small mammals together constitute approximately another 25% of households. This statistic is fairly uniform across European states.

There seems to be universal agreement that pets are of therapeutic value. Studies have shown (Wells 2011-review) that many human disorders, including behavioural, emotional and physical conditions, have improved with regular contact with pets, particularly dogs or cats. There are many cases where animals (typically dogs) provide actual physical assistance; the most well-known example is the guide dog for blind people. Increasingly dogs are being used for hearing, detecting medical conditions and general assistance. Whilst there main role may be utilitarian, there is no doubt that the animals frequently fulfil an emotional role for their disadvantaged owner.

Pet owning may be of particular value in urban populations where density and lifestyle create stress. However, it is just these urban situations that make pet-keeping more difficult. Provisions that facilitate pet-keeping often overlap with nature-promoting design of the NEAR and FAR environments. The provision of a secure outdoor space and access to it may be a critical factor in cat or dog ownership and the provision of space for a pond, rabbit cages or aviary similarly important. Dog-walking, an almost essential part of dog ownership, has needs in terms of access to FAR nature environments, but also has great benefits since it in itself promotes health through exercise as well as exposing the walker to nature and ensuring that the nature provision is well used.

Pet-keeping can create a number of local inter-neighbour problems. These usually centre around noise, hygiene and perceived danger. To some extent these are design issues, some of which such as noise, already being covered by codes of practice. Some are generated by anti-social and thoughtless behaviour, and cannot be designed out.

Though these seem highly specific issues, outside the normal architectural design sphere, bearing in mind the prevalence of pet-keeping, and its benefits to health and well-being, it seems appropriate to bring some design effort to bear to this neglected area.

4 Conclusions

The literature provides convincing evidence that access to nature has a significant impact on health and well-being; only a tiny fraction has been quoted here. It follows that it should be an essential part of the design and planning of residential buildings. This paper suggests a framework for how a design proposal could be assessed. No quantitative calibration has been offered yet. But by means of social survey and Post Occupancy Evaluation, a quantitative model could evolve.

References

Chalfont, G.E.: Design for Nature in Dementia Care. Jessica Kingsley Publishers, London (2008)

English Nature: Wildlife-Friendly Gardening – A General Guide (2005)

Faber Taylor, A., Kuo, F.E., Sullivan, W.C.: Views of nature and self-discipline: evidence from inner city children. J. Environ. Psychol. **22**, 49–63 (2002)

Fjørtoft, I.: Landscape as playscape: the effects of natural environments on children's play and motor development. Child. Youth Environ. **14**(2), 21–44 (2004)

Grahn, P., Martensson, F., Lindblad, B., Nilsson, P., Ekman, A.: *Sveriges lantbruksuniversitet, Alnarp.* UTE pa DAGIS, Stad & Land nr. 93/1991/1997 (1997)

Greater London Authority: Improving Londoners' Access to Nature London Plan (2008)

Hellinga, H.I.: Daylight and View; The Influence of Windows on the Visual Quality of Indoor Spaces. PhD Thesis, University of Delft (2013)

Heschong Mahone Group: Windows and Offices: A Study of Office Worker Performance and the Indoor Environment. Technical Report for the California Energy Commission (2003)

Kaplan, R.: The role of nature in the context of the workplace. Landsc. Urban Plan. **26**(1–4), 193–201 (1993)

Kaplan, R.: The nature of the view from home: psychological benefits. Environ. Behav. **33**(4), 507–542 (2001)
Markus, T.A.: The function of windows – a reappraisal. Build. Sci. **2**(2), 97–121 (1967)
Nicol, F., Humphreys, M.: Thermal comfort as part of a self-regulating system. Build. Res. Pract. **1**(3) (1973)
Strife, S., Downey, L.: Childhood development and access to nature. Organ Environ. **22**(1), 99–122 (2009)
Ulrich, R.S.: View through window may influence recovery from surgery. Science 224, 420 (1984, April 27)
Wells, N.: At home with nature, effects of "Greenness" on children's cognitive functioning. Environ. Behav. **32**(6), 775–795 (2000)
Wells, D.: The value of pets for human health. Psychologist. **24**, 172–176 (2011)
Wilson, E.O.: Biophilia. Harvard University Press, Cambridge, MA (1984). ISBN:0-674-07442-4

Virtual and Augmented Reality Applications in Building Industry

Shahrokh Shamalinia

Abstract Virtual reality (VR) and augmented reality (AR) technologies have been emerging rapidly in the past few years and are expected to redefine the way we interact, communicate and work together. It appears that this emergence is in perfect timing with the growth of Internet of things (IoT) devices and can form a killer combination to go beyond the limits of what is possible today in different industries including architecture and the built environment. In this short introduction, we will look at the evolution of smart devices and how they are expected to shape the future of our connected smart cities. We will finish with suggestions for use case scenarios of these technologies applicable to different areas of the building industry.

1 The Need and the Distance

Architecture has always been a response to human needs. These needs have varied through ages, but some of them have been there for the entire life of humans such as the need for protection from natural forces or predators (idea of shelter.) Satisfying more intricate needs of the occupants of a building of any kind and purpose has always been a bigger challenge for the architect.

Although as a hunter-gatherer we were not feeling the need for a permanent shelter (it would be useless maybe since we were always on the move) after settling down and change of lifestyle to a more stationary one, the need for a permanent refuge that was capable of responding to the new social life was eminent (Harman 1999).

With the technological advancements of our age have come a great number of new needs. Our life has changed drastically by computers. We don't necessarily need to move from a village to another in order to deliver a message. We manage to do many things without even moving. We can even work from home! And the fun part is that when we have to move, we can take our little computer friends

S. Shamalinia (✉)
Independent Researcher, Tehran, Iran
e-mail: shahrokh@3dbryx.com; sh_alinia@yahoo.com
Website: www.3dbryx.com

S.T. Rassia, P.M. Pardalos (eds.), *Smart City Networks*, Springer Optimization and Its Applications 125, DOI 10.1007/978-3-319-61313-0_2

(smartphones) with us. We carry them nicely in our pocket, and as soon as we reach the destination, we choose a table and maybe the first thing we do is to check our phone or tablet.

The core of our interest in smart connected devices is to access the collective human knowledge and information, which in turn has taken the shape of a basic need of our digital age. A connected device is a tunnel that most probably will lead us to an answer whether it is the current temperature or latest news or translation of a word in a language that we don't understand. This is how far the hunter-gatherer has reached, and obviously there will be no stopping here.

Telephones used to be devices that connected us to the person on the other side of the line. They were bulky, stationary, and wired. Later we invented mobile phones which we could use to perform the same tasks on the go. But they were not connected to the World Wide Web. Today many people carry a smartphone that can connect to their home security system or log in to their office computer while attending a meeting abroad. New "things" are emerging capable of connecting to you or to a network you can join. They are also growing in number and functionality as well.

Our mobile phones have left their stationary ancestors and managed to walk with us everywhere. We realize that with their elevated importance and functionality of their newer generation (the smartphones), we keep them even closer to us. Some may even fall asleep while using their phones in the bed. Many others wake up to a sound notification of an incoming message on their phone or check their social media updates as the first thing they do after waking up. This is an important level of intimacy to take note of, but it is also important to understand that we will continue to draw our smart tools even closer (Fig. 1).

Has the physical world turned into a mere interface that we use in order to facilitate our access to the digital world? Why allocating only a few inches of screen space to something that has such great value for us? What if we could get them much closer to our eyes so that it can cover all of our field of view? This is exactly the case with VR and AR technologies that we are going to discuss in the next section.

2 Tools

We will be introducing two main groups of tools that can be used in building industry: virtual reality (VR) and augmented reality (AR).

2.1 VR Devices

VR technologies currently fall into two main categories: mobile-based and desktop-based head-mounted displays (HMDs). The first group is compatible with smartphones (Fig. 2a) and gives you a 360° view to a new form of media. It

Fig. 1 The distance between us and our smart devices is decreasing. Will we be wearing our smart lenses to find ourselves in a virtual world simulated around us (No. 5) or even take a pill that performs at our neurons level to reconstruct a whole new set of realities for us (No. 6)? (© Shahrokh Shamalinia)

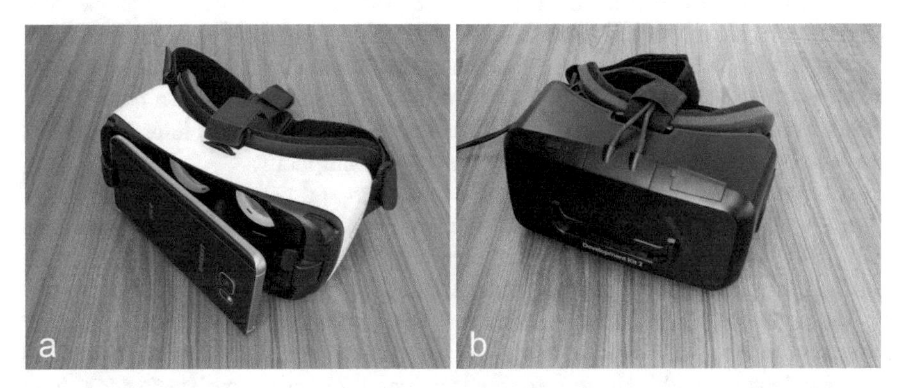

Fig. 2 (**a**) Samsung Gear VR with a compatible phone; (**b**) Oculus Rift DK2 HMD (© Shahrokh Shamalinia)

consists of a viewer with two lenses and a place to insert a mobile phone that runs the actual VR application. So basically you slide your smartphone into a vertical slot on the HMD and keep it close to your eyes (or wear it) such that you will be looking at your phone screen through a couple of lenses that adjusts your view and creates the illusion of three-dimensional depth similar to the way our eyes work. This is a fantastic solution since it is light-weight and wireless. However, it lacks the head tracking functionality that the latter group offers. The desktop-based HMDs (Fig. 2b) include a dedicated front panel high-resolution screen and need to be wired to a desktop computer. They also come with a special infrared camera that keeps track of the user's head location and orientation (head tracking), which leads to a much more immersive VR experience compared to mobile-based HMDs with the expense of less mobility.

Both types of VR are designed to give you the total immersion by taking you to the center of a completely different world than physical reality. As an example you can pop into a 3D prehistoric cave recreated by VR developers (or maybe 3D scanned from the actual place) and examine the paintings on the walls all around you, or you can simply switch to another VR experience in the depth of an ocean and learn about different kinds of fish without actually going down there!

2.2 AR Devices

There are two other forward-looking technologies that are gaining more and more attention these days: augmented reality (AR) and mixed reality (MR), both of which augment your surrounding physical world with 3D virtual content depending on the application you choose to interact with. For example, you can run an AR application on your phone and point it to a specific image marker on a magazine to show additional augmented content such as a video or a 3D object on your phone. MR applications are a hybrid of both VR and AR where you would wear an HMD, but you will still be able to see the physical world around you through the phone's camera and still have the 3D virtual objects and infographics spawn around you. Since both AR and MR are very similar and are being used almost interchangeably, we will be referring to them as AR for the sake of simplicity.

Currently AR technology is available both with and without special glasses. A majority of AR applications are built to run on mobile phones. The following image (Fig. 3) shows an AR application augmenting 3D model of the building when the phone camera detects the specified marker image.

Another and perhaps more serious way of using AR is to wear special glasses that covers the user's field of view and brings augmented content on top of the visible physical world. Using these devices, you will have complete freedom over the interaction since there is no need to hold and control the device with your hand.

3 Use Cases

These technological advancements have great potentials to shape our day-to-day interactions and facilitate access to the big data in a city network or even a global one intuitively. As an example there is already a big data on weather being crunched constantly every hour, but still those numbers need to have a visual interface to make

Fig. 3 Example of an AR application augmenting 3D model of a residential area on a piece of paper (© Shahrokh Shamalinia)

any sense for a given user. What if we had glasses that could use the weather data and overlay it on top of what we see through them, showing us how the weather will look like in this specific GPS location a week or a month from now using voice commands? What if urban designers could wear those glasses to evaluate the efficiency of the built environment by just looking at the buildings with overlay color spectrums that highlight the undesirable corners in red? Solarchvision[1] has been developing a computer program that analyzes the sun direction to produce a heat map based on factors of desirability for a given set of 3D buildings (Samimi 2015). All there is left to achieve the abovementioned scenario would be to overlay that information on top of an AR application for wearable devices (Fig. 4).

It is imaginable for every building to have a virtual tag that once scanned through your glasses (or even AR-enabled contact lenses in future), it will visualize key information about that particular building and how to use it such as access to emergency exits or the fire escape or a pathfinding line that is virtually overlaid on the floor to take you to the room or person you are looking for.

All these require the buildings and the cities to have a network of connected smart devices. The Internet of things (IoT) will play a pivotal role in the way we interact and live in the cities.

Here we will briefly introduce cases where VR and AR technologies can help building industry to achieve better and more efficient results in areas such as design, construction, planning, maintenance, presentation, and education.

[1] www.solarchvision.com

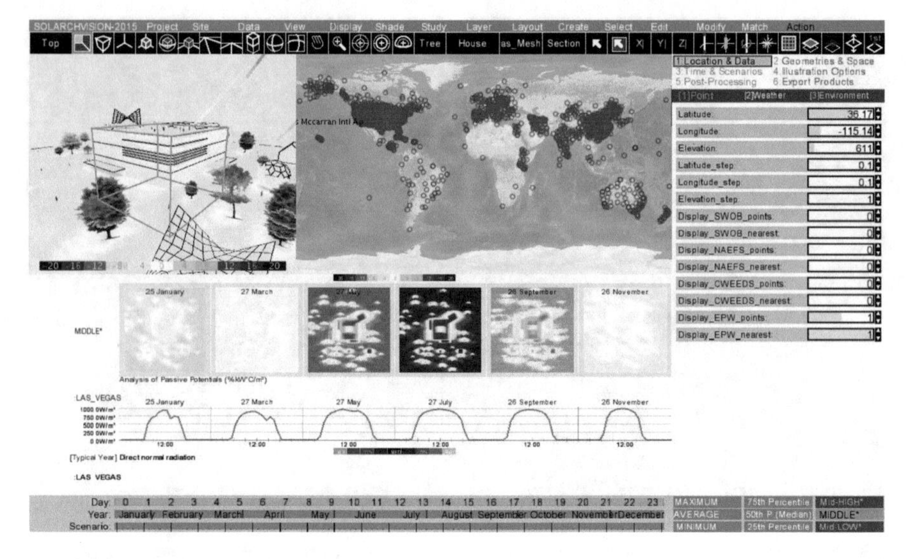

Fig. 4 Heat maps created using solar information for a set of 3D models in a specific time and location (© SOLARCHVISION)

3.1 Design

When designing a building using CAD software, the architect works with plans, sections, elevations, and the resulting 3D model of the building. These have been great interfaces in assisting architects to shape their imagination before the actual building is completed. Of course there have always been limits to what we can imagine through 2D plans, elevations, and sections. Even best 3D visualizations of the buildings cannot replace the feeling of actually being surrounded in the unbuilt project.

Using VR tools, the architects will now be taken right into the virtual building with the correct scale, materials, textures, reflections, and real-time lighting, all happening around them rather than on a 2D monitor. This exceptional experience is a new phenomenon in the history of architectural design. For the first time in ages, we are going to walk through the unbuilt and interact with it completely free from physical limits of matter and gravity (Figs. 5 and 6).

AR also offers great options for a designer. Wearing an AR glass loaded with a proper design application, we can begin creating virtual objects that fit right into the immediate physical world around us. Imagine an interior design or a renovation project where the designer can augment parts of the future design onto the existing structure and share that with his team in real time. The AR-enabled design software can have access to real-time solar and weather data for the specific location which can be called into augmented view to give the architect additional layers of information highlighting walls or floors that have the most optimum design based on these environmental factors.

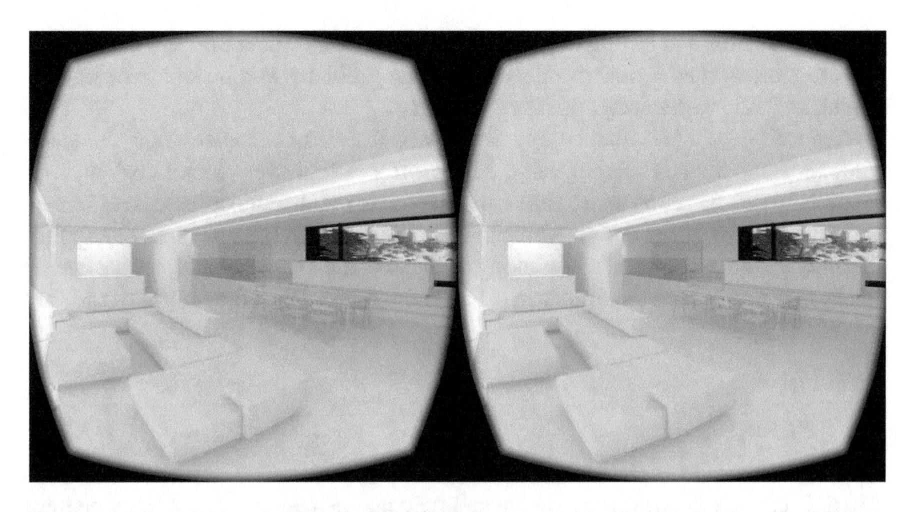

Fig. 5 A screen capture of an Oculus Rift PC application showing inside of a virtual building (© Shahrokh Shamalinia)

Fig. 6 A screen capture of a Gear VR Android application showing exterior of a virtual building (© Shahrokh Shamalinia)

3.2 Construction

On-site inspection with AR glasses will probably be the most intuitive way to assure accuracy and safety of a construction site. Wearing one of these glasses, you will be able to see which beam or column needs to be rechecked or if the drawing specifications have been correctly applied to the built structure. On-screen alerts can bring a faulty unsafe connection into attention immediately preventing potential hazards and saving lives. Text and video instructions can be overlaid on a construction worker's view to safely rectify the problems and make easier smarter decisions while under such critical conditions.

According to MIT Technology Review website, one construction company (Gilbane) has already started using Microsoft's AR glasses (HoloLens) in their process. In one occasion using HoloLens, they realized that the ordered steel frames were too long to fit the design which gained them enough time to "ask the supplier to cut the frames shorter in his shop rather than make workers adjust dozens of tracks that would hold the frames in place," which saved the company about $5,000 in costs.[2]

3.3 Planning

Big data created in cities are massive and they are growing instantly. They contain many different aspects of measureable factors that affect our everyday life. City planners use these data to design, manage, and maintain the city sustainably. Virtual 3D models that are fed with these data can make a very close simulation sandbox for planners to examine the behavior of different aspects of the system in a safe virtual environment.

The city of Singapore has recently created a similar project called Virtual Singapore. An exact replica of all the city buildings is created together with artificial intelligence that simulates the crowd behavior and real-time physics. This can be a tremendous tool to help decision makers and city planners to understand and solve complex problems working with 3D visible outcome of the big data rather than spread sheets filled with numbers and text information.

> Cities are some of the most complex 'products' created by humanity. Through more efficient and accurate predictions of future experiences within these cities using state-of-the-art tools and applications, we can better anticipate national resource planning or provision of services, and contribute towards a more sustainable quality of life. We hope to see other cities echo Singapore's exciting initiative.[3] (said Bernard Charlès, President & CEO, Dassault Systèmes)

3.4 Maintenance

There are many layers of embedded electrical, piping, and structural elements woven into the fabric of a typical building, most of which are not visible when there is a need to repair or maintain them. The only way to find out the exact location of electrical or data cables is to consult the building blueprints.

[2]https://www.technologyreview.com/s/602124/augmented-reality-could-speed-up-construction-projects/

[3]http://perspectives.3ds.com/architecture-engineering-construction/virtual-singapore-a-platform-to-solve-emerging-and-complex-challenges/

With the aid of the AR-enabled glasses, it is now possible to overlay these hidden elements on top of the actual physical walls and floors. This way by just looking at the building, we can find out where a faulty pipe is located or what the best approach to replace a data cable is. A simple customizable application can monitor these elements and alert the habitants even before a service failure is about to happen giving them adequate time for preparation.

3.5 Presentation

Traditional presentation methods in architecture usually consist of rendered images, animations, drawings, and some text information composed into one or multiple sheets. They have worked great so far, but there have also been shortcomings. Only a selected set of views are presented while a building is a spatial product that needs to be experienced rather than imagined through pictures.

With VR technology today, we can take the client inside an unbuilt design and give them the option of freely moving through the space experiencing the visuals, scale, sounds, distances, lighting, and all elements that define an architectural space by themselves. This can also open up doors for better and more interactive feedbacks from clients since they can be given the tools to virtually change the colors, size, or placement of objects to their likings in real time and share it with the architect in 3D for faster and more efficient communication than just putting them in words.

Another great traditional presentation tool has so far been scale models. 3D models can never compete with them as long as they are shown on monitors, but what if we bring them out of the constraints of a 2D screen and put them on the meeting table (Figs. 7 and 8)? We can now do that with the help of AR. A simple AR application loaded into a mobile phone can place the 3D model of a building on a printed target image. The physical paper and virtual model are then joined together such that by moving the paper, the 3D virtual building will also move with it. Now everyone in the meeting room with a smartphone can download and install the application and point it to the printed paper to see the attached virtual building on it. They can choose to make the model smaller or bigger in size or even change the colors or show/hide different embedded layers of information from the screen of their smartphones and save a copy of those modifications for themselves or to share with others. Obviously a traditional scale model will not be capable of these new functions and will remain limited to the constraints of the physical world. No need to mention that the environmental costs of physical models and the recycling challenges of their materials after use would be another reason to encourage use of digital solutions for architectural presentations.

Finally, there are many cases where new updates or changes to the design require the whole physical scale model to be rebuilt, which inevitably result in higher costs and loss of valuable time. Using AR models will eliminate these costs while the updates are pushed into a newer version of the application that can be downloaded in a matter of seconds on the clients’ phones.

Fig. 7 An AR application augmenting a 3D-scanned interior of a historic building on top of a card (© Shahrokh Shamalinia)

Fig. 8 Why make physical scale models when you can augment your 3D design on a printed plan? (© Shahrokh Shamalinia)

3.6 *Education*

Here I would like to present a few scenarios where architectural education may hugely benefit from the new technologies we have discussed so far. Although some of them may seem so far-fetched, we will still need to let our imagination fly as high as possible for possible future inventions.

3.6.1 VR Studios

Architecture studios and workshops traditionally have been rooms with big drawing tables and huge windows to bring the natural light in since students and professors would physically need to attend the class and use the space. They may also be equipped with computers and video projectors to bring additional educational material to the class.

A VR studio is a virtual reality space that can be logged in using a VR HMD and accessed from anywhere with an Internet connection. There will be no need for a physical video projection inside a VR studio since any digital content can be presented right into the HMDs of individual participants. With a seamless user interface design, everyone will be able to communicate with other classmates' avatar as well as receive text, image, video, or 3D feed at the corner of their visual periphery on the HMDs. Instead of taking a laptop to the workshop, students will be taking the workshop into their HMDs where they can pull up a 2D browser and a virtual keyboard at any moment using their virtual finger tips.

These may sound like science fiction to many of us (excluding Ernest Cline the great author of *Ready Player One* novel) (Cline 2011), but today we already have virtual social networks and hand-tracking devices that work with HMDs to manipulate virtual objects and even sculpt in stereoscopic 3D. Oculus has introduced Medium, an application designed to give you sculpting powers in virtual reality.

Once again we are able to build with our hands, but this time without even feeling the weight of a single brick! But wait, will this pose a problem if we take gravity out of the equation? We will address that on Virtual Building Simulators section below.

3.6.2 Virtual Building Simulators

Architecture students have always learned to pay close attention to structural and mechanical aspects of the building parallel to the design and aesthetics. However they learn those disciplines through textbooks and engineering manuals and barely have a chance to experience the physics behind them. Of course it may be life-threatening to experience an overloaded beam or structural deformity in a real-world earthquake, but in a VR simulation, you can always remove the HMD or simply close your eyes if things start falling!

A building simulator can take your design in 3D and run on-demand simulations in real time while you see the model right in front of you or while you are standing in it. You can record, play, pause, and rewind the simulation for further inspection and get feedback from the application's built-in artificial intelligence (or your human instructor) as to where the problem is and how to fix it (more on this in Virtual Design Assistants section below). Gamification of such training materials can make it more engaging and maybe more fun to play with to efficiently learn abstract engineering concepts in virtual practice.

Virtual building simulators may not be limited to engineering and mechanical aspects of a building, but also they can target the social and psychological aspects involved. With the advancement of the artificial intelligence, we can expect to be able to simulate human-level behavioral patterns in the future. A future that can only be as far as 2040 according to Ray Kurzweil:

> The other approach is to take one or more biological human brains that have already gained sufficient knowledge to converse in meaningful language and to otherwise behave in a mature manner and copy their neocortical patterns into the simulated brain...I would not expect such an "uploading" technology to be available until around the 2040s. (Chapter 7) (Kurzweil 2012)

Will this mean we can spawn multiple instances of these AI entities that represent individual human beings with different social and cultural backgrounds uploaded to them and evaluate their reaction to the design of a proposed building before it is built? Will architects and urban designers be able to effectively evaluate the effectiveness of their design proposals and how the simulated crowd will be responding to them in virtual worlds? Westworld, a science fiction movie made in 1973, was one of the pioneers of this dream followed by a recent and more detailed TV series with the same title from HBO TV that illustrates the complexity of bringing artificial intelligence, augmented reality, virtual reality, and immersive story-telling in one place.

3.6.3 Virtual Design Assistants

Many of the parameters involved in a successful architectural design in elementary levels are programmable into a set of basic rules that can be monitored with artificial intelligence. Topics such as size, scale, accessibility, floor heights, ramp slopes, sun orientation, energy efficiency, choice of materials for the specified geographical location, and many more that fit into quantitative aspects of a design can be checked for accuracy with any given 3D design.

The key idea here is to connect tons and tons of guidelines and technical architecture textbooks available in our collective treasury of the Internet, with the end users (i.e., students) in a unified visual and interactive way that otherwise would be missed due to the vastness of the available data. Having a personal design assistant that you can carry in your pocket or even wear as smart contact lenses that overlays information on your field of view may just transform the way you design and research in the future.

3.6.4 GEO AR Applications

Imagine a typical hotel building that is going to be built in a specific location in the city. This can be a student project or even an architectural competition to elect the best design. Currently the closest we can get to place the unbuilt building in the site is to photo-montage a rendered image of the building on top of an actual photograph taken from the same vantage point. But as we discussed before, this will just be too limited and frozen in time. Architecture is experienced through movement, just like a movie. Experiencing architecture through still images can be compared to watching a movie through single still frames as opposed to 30 frames per second.

GEO AR Applications help us to overcome this problem by placing the virtual 3D design alternatives of the building right into the geo-located space. Everyone wearing AR-enabled glasses, lenses, or smartphones can see the virtual building in place before it is built. Moreover, they can interact with the application to pull up the next alternative design and vote for them making it very easy for the decision makers, professors, or the jury to get the public opinion on each design.

Our hotel building example will be checked against many factors: Is it blocking an important view from other side of the street? Is it getting enough sunlight in the winter? How will the material of the façades work with the existing urban context? To find out about these questions, all we will need to do is to take the phone and examine the virtual building alternatives directly.

4 Conclusion

We seem to be standing at a very decisive point in time where the new AR and VR technologies are beginning to emerge and transform the way we interact with the built environment. Moreover, the infrastructures to make smarter connected cities are growing rapidly with new IoT devices emerging with the same pace. While these technologies have great potentials to introduce efficiency and accuracy for the building industry as well as facilitating design communication and presentation for architects, there may still be some questions to think about.

Will the world of architecture make use of these advancements any time soon? Will architects and urban designers prefer the new VR and AR tools over the traditional ones? Will architects still rely on 2D plans, sections, and elevations as first steps in designing a new building as opposed to designing in 360° 3D interactive virtual space?

Acknowledgment The author wishes to acknowledge the contribution of Mojtaba Samimi by sharing his wisdom and insights in initial ideation phases of writing this article.

References

Cline, E.: Ready Player One. Crown Publisher, New York (2011)
Harman, C.: A People's History of the World. Bookmarks Publications, London (1999)
Kurzweil, R.: How to Create a Mind, the Secret of Human Thought Revealed. Viking Penguin, New York (2012) PDF
Samimi, M.: Weather data and solar orientation. In: Rassia, S.T., Pardalos, P.M. (eds.) Future City Architecture for Optimal Living, Springer Optimization and Its Applications, pp. 221–240. Springer, Cham (2015)

Asia's Cities: Necessity, Challenges and Solutions for Going 'Smart'

Rupesh S. Iyengar

Abstract Asia is developing, and its cities are going to play a major role in this endeavour to match developed counterparts. Asian trade, population, geographic size of its cities and contribution to global development will only increase in the years to come. Rural settlements or underdeveloped villages are fast converting themselves to smaller towns; smaller towns are converting themselves into small cities, and existing small cities are forging ahead into becoming megacities. This demographic transformation in the urban landscape will only increase the use of resources like land, water, clean air, sanitation, power, transport network and safety in order to survive and grow. The quantity and quality of investment that Asian cities make today in these resources will help them service and sustain their burgeoning population in the future. It is therefore imperative that urban planning, use of technology, futuristic vision and control techniques that are incorporated, work in collaboration to achieve success. Present-day megacities like Tokyo, Seoul, Beijing, Shanghai, Manila, Jakarta, Mumbai, Delhi, Karachi, Istanbul, Tehran, Moscow, etc. have their own share of problems; and negotiating their population's ever-growing demands have turned into a herculean task. These cities can boast of a glorious and historic past, but how they mitigate their current issues, envisage future needs and meticulously plan their future are important. The concept of smart cities is not to be misunderstood as only constructing idealistic new cities from the scratch. While this could be constituted in some developed economies in Asia, having existing megacities and their urban sprawl change their style of operation to suit present and future needs could be a smarter and more beneficial solution. Cities are never built in a single day; they always evolve with time and with the evolving cultural fabric of its residents.

R.S. Iyengar (✉)
Services Consultants, 1306/110, 29th B Cross, 13th Main, 4th Block East, Jayanagar, Bengaluru 560011, India
e-mail: rupeshiyengar@gmail.com

S.T. Rassia, P.M. Pardalos (eds.), *Smart City Networks*, Springer Optimization and Its Applications 125, DOI 10.1007/978-3-319-61313-0_3

1 Introduction

Asia is developing, and its cities are going to play a major role in this endeavour to match developed counterparts. Asian trade, population, geographic size of its cities and contribution to global development will only increase in the years to come. Rural settlements or underdeveloped villages are fast converting themselves to smaller towns; smaller towns are converting themselves into small cities, and existing small cities are forging ahead into becoming megacities. This demographic transformation in the urban landscape will only increase the use of resources like land, water, clean air, sanitation, power, transport network and safety in order to survive and grow. The quantity and quality of investment that Asian cities make today in these resources will help them service and sustain their burgeoning population in the future. It is therefore imperative that urban planning, use of technology, futuristic vision and control techniques that are incorporated, work in collaboration to achieve success. Present-day megacities like Tokyo, Seoul, Beijing, Shanghai, Manila, Jakarta, Mumbai, Delhi, Karachi, Istanbul, Tehran, Moscow, etc. have their own share of problems; and negotiating their population's ever-growing demands have turned into a herculean task. These cities can boast of a glorious and historic past, but how they mitigate their current issues, envisage future needs and meticulously plan their future are important. The concept of smart cities is not to be misunderstood as only constructing idealistic new cities from the scratch. While this could be constituted in some developed economies in Asia, having existing megacities and their urban sprawl change their style of operation to suit present and future needs could be a smarter and more beneficial solution. Cities are never built in a single day; they always evolve with time and with the evolving cultural fabric of its residents.

2 Development of Asian Cities

Asian cities have had long recorded histories. Human civilisation and large urban dwellings thrived in Asia since 2200 B.C., from the times of 'the Harappan civilisation'. From the kingdom of Angkor in today's Siem Reap, Cambodia; cities of China during the Han, Xie and Ming dynasties that started and developed around Beijing; and the kingdoms of Mysore and Rajputana in Southern and Northern India, Asia is witness to the rise and fall of megacities over centuries of time. More than two thirds of present Asian cities were already important cities since 500 years ago, and more than a quarter of them were founded more than 2000 years ago. Asia has had most of the world's urban population and most of the world's largest cities. During the mid-1960s, only 20% of Asia's population was urban and lived in cities. Today that number stands at 48% (United Nations 2014a). This steep rise in urban diaspora has exceeded all expectations and has made redundant all calculations. It is projected that almost 64% of the Asian population will be living in cities by 2050

(United Nations 2014b). That accounts to more than 40 million people being added to Asia's urban landscape every year. This trend, if not being constant in the years to come, is only expected to rise. Cities in Europe and America that are classified as urban cities have a population size of five to ten million people. In contrast, many Asian cities have 10–20 million people. They therefore are rightly classified as megacities today. Statistics reveal that 14 among the 25 megacities of the world today are in Asia with many of them in India and China. Moreover, second-tier cities of India and China would assume population proportions of urban cities in the west. The urban centres of Asia are therefore home to around 1.8 billion people, i.e. a quarter of the world's population and about half of its urban population. By 2025, around a third of the world's total population is likely to live in Asia's urban homes.

Most Asian countries are developing economies as classified by the World Bank. That means that many of these countries have a large percentage of their population living below the poverty line in slums and squatter settlements among urban congregations. This section of the urban population lacks the minimum requirement of water, drainage, sanitation, quality healthcare and educational facilities. Indications that such infrastructure are being razed will hint towards reducing poverty and meeting of international development targets such as the Millennium Development Goals (Poverty 2015). Asia's percentage of world population, concentration of urban population and size of urban cities and megacities reflect the region's large and increasing role in world economy. Asian cities are burgeoning business destinations and contain a sizable part of all domestic and foreign investments and trade. Asia is home to the world's seven largest economies of the top 20 and has the second, third and fourth largest economies, namely, China, India and Japan, respectively. Most countries in Asia now have more than half their GDP from industry and services sectors that have their businesses in urban areas. In general, the theory is that, higher a nation's per capita income, higher is the percentage of urbanised population. Also, the more rapid the country's economic growth, the greater is the increase in its proportion of population living in urban cities. So with economic growth, the burden on existing cities of Asia will only grow with time. Table 1 gives information on million plus populous cities of Asia according to the United Nation Report (United Nations 2014a).

The importance of Asian cities is thus evident, thanks to the rapid transformation of its cities into megacities. Megacities like Tokyo and Shanghai are now becoming the hub of economic activity and economic growth. World businesses will continue to increase and will have to pass through these megacities. Interdependency of western cities on their eastern counterparts for trade and business will only increase with time. Outsourcing of businesses by the west to cities in the east is rising, and the opportunities in industry, commerce, services, entrepreneurship, innovations, healthcare, hospitality, education, research, infrastructure, living spaces, etc. are only increasing. Some cities like Dubai, Singapore, Bengaluru and Manila have already started to play pivotal roles for transit and travel, jobs and employment, technology and communication, healthcare and infrastructure, etc. The main challenge for Asian cities, however, is the scale and pace of urbanisation. The key to a structured form of urban planning is to have, firstly an integrated system of

Table 1 Distribution of Asia's largest cities and their million plus population in 2014

Nations listed by the size of their GDP (PPP) in 2016	No. of cities with 1–5 million population	No. of cities with 5–10 million population	No. of megacities with more than 10 million population
China	122	12	10
India	54	4	3
Japan	13	0	1
Russia	10	1	1
Indonesia	14	0	1
South Korea	8	0	1
Saudi Arabia	4	1	0
Turkey	5	1	1
Iran	7	1	0
Thailand	0	1	1
Pakistan	8	0	2
Philippines	4	0	1
Bangladesh	2	0	1

infrastructure, second an interactive urban transport network and finally a model for zoning of developments. India was predominantly an agricultural economy. But since the last three decades, agriculture has been uprooted by industrial and service-oriented economic sectors. Similar is the growth story of China, Indonesia, Philippines and some of the other developing economies of Southeast Asia. Changing lifestyles and adaptation to different cultural ecosystems could be the reason for this urban transformation and mushrooming.

Urbanisation, if done in a structured manner, could increase lifestyle standards, reduce poverty, achieve sustainable development, maximise economic efficiency and ascertain the good practice of urban living. The criteria for sustaining an urban ecosystem are good governance, improved urban management, effective and efficient infrastructure and service provision, financing and cost recovery, social and environmental sustainability, innovation and change and leveraging international development assistance. Policymakers need to envisage the city's growth and have a vision to promote safe, liveable, well-managed and environmentally friendly spaces that are free of poverty. Funding needs of cities' infrastructure should be adequately provided to have an integrated approach to combining energy, water, ventilation, reuse and recycling of resources. Promoting a low-carbon lifestyle and the idea of low-carbon cities that use fewer inputs of water, energy and food and simultaneously produce fewer outputs of waste, heat, air and water need to be encouraged. Ecological impacts on natural resources, forests and original habitations should be minimised by providing a healthy, safe and comfortable environment to its inhabitants. Energy conservation measures and use of renewable energy sources like hydrogen power, fuel cells and biofuels should be encouraged instead of using fossil fuels. Cities should be developed beyond the needs of work or commercial living and enjoyment but to make them complete in all means so that those who live in them feel that they are self-sufficient, empowered, safe and enjoy a clean environment. The western world is therefore inquisitive to learn how these burgeoning megacities of Asia would function, survive and serve their populations in the future, as their development has a large impact on the rest of the world.

3 Megacities of India and China: Current Challenges

Chengdu was once a lethargic but lush city in Central China with a population of 500,000 in the 1950s. Today the same city has been transformed into a bustling megapolis of 14 million people (Xueqin 2012). Moving from the semirural to a megacity ecosystem has brought about many changes in lifestyle, culture and taste of Chengdu's urban population. While some people welcome this transformation that has improved livelihood and purchasing power, others just dwell in the mirage of lush gardens and pure air that once were in abundance in this urban dwelling of Central China. Chengdu had been known and was famous for its tea houses, where people used to pass time playing cards in the evenings. But today the city boasts of two new monuments aimed at bringing the global spotlight back onto this city. The New Century Global Centre, regarded as the largest single building in the world, is a leisure complex that houses two 1000-room five-star hotels, an ice rink, a luxury Imax cinema, vast shopping malls and a 20,000-capacity indoor swimming pool with 400 m of 'coastline' and a fake beach, which is the size of 10 football pitches complete with its own seaside village. Alongside there is another massive and futuristic structure, a contemporary arts centre designed by the award-winning Iraqi-born architect, Zaha Hadid.

Megacities may be overwhelming, but they come with their share of problems and at a price. The centre of Johannesburg, a recently minted megacity, is known for its slums of frustration and desperation, while its circumference grows big with financial centres and gated communities. Beijing, which was once the seat of the Forbidden City and heart of many famous Chinese dynasties, has a landscape that has been transformed by tall skyscrappers and lengthy highways. The by-product of Beijing's overwhelming response to becoming a megacity has resulted in overcrowding of urban dwellings, traffic congestions on its 20-lane highways and, more recently, a serious air pollution problem. The unquenchable appetites of Beijing for fuel, energy and water have depleted the surrounding provinces of their own potential. The problem usually lies in the uncontrollable influx of migrants from other parts of the country making the management of resources and amenities a daunting task for the city administration to handle. Cheap labour, increased monetary benefits, high standard of living and better healthcare are some of the primary objectives that cause this migration. The main problem with Chengdu's growth and other urban centres of China is the mentality with which the growth is driven. Sometimes it is referred to as 'growth for growth's sake', which emphasises buildings and statistics over people and ideas. China's city planners need to understand that a city exists to unite and inspire its people and to engage in creative endeavours that would better themselves and their city and not merely for the sake of building facilities aping some other city in the west. Cities have to be organic and dynamic, and they have to inspire its citizens and diverse communities to thrive there. Cities need to be creative and form an integrated urban dwelling that has technological, economic, cultural, social and artistic fabrics intertwined with each other. Physical attractions in cities like sports stadiums, highways, urban malls, parks, tourism and entertainment centres become irrelevant and insufficient and merely attain the status of unattractive and boring landscapes with time if there is no strong history or creativity or cultural emotion attached to it. What makes some urban cities like Paris or Berlin to stand apart from the rest is this deep sense of creativity or history that is attached to them. Openness to diversity of all kinds and above all the opportunity for creative minds to validate their identities are what makes urban ecosystems flourish. Chengdu, a land parcel known for its beautiful rural hills, distinctive culture, traditional openness and tolerance, artistic communities, etc., could have become China's top creative centre if it was built on its strengths of multicultural diverse living. Ironically, the city planners chose rapid and culturally asynchronous urbanisation. Chengdu thus is losing its identity and character. It could be said that it is moving towards becoming a little sister of Beijing where a tense conflict between buildings and people has alienated one from the other. It is time that cities are designed with more creative and thoughtful visions and not by eroding the glorious cultural past of the city.

The problem in India however stems out of the sheer poverty that is prevalent in its megacities. The rural economy is not enough for inhabitants to have a minimum standard of living and absence of quality education, healthcare, high-paying jobs and luxuries incentivise people to make the move to cities. The cities in turn are only planned for a certain population size. The rate of population import into the cities do not match the urban planning measures taken up by the city's administration in developing facilities, infrastructure and dwelling communities. This in turn gives rise to unplanned habitation of land as and where it is available. Much of the present growth in India's megacities take place by building of shanty makeshift temporary homes or slums, popularly known as 'bastis' on a piece of land that happens to be available and unoccupied. The main challenge for Indian cities is the exacerbated population growth and rapid urbanisation combined. According to McKinsey, the country's cities are expected to grow from 340 million people in 2008 to a whopping 590 million in 2030 (Sankhe and Vittal 2010). Meeting demand for urban services in these cities will require US$1.1 trillion in capital investment over the next 20 years (Sankhe and Vittal 2010). Absence of the right design and planning could only mushroom existing problems of traffic safety, congestion and pollution in the years to come. Some of the main obstacles limiting India from having a structured urban landscape development are:

1. Urban Slouch

Since the last two decades, cities of India have grown not only in population but also in geographic proportion. For example, In the last 20 years, Delhi's urban footprint has increased two fold. This in turn has led to many changes in commuting.

The average trip length has increased from 8.4 to 10.5 km, and this phenomenon is only set to increase further. The sprawls of modern cities and rise in commuting distance have increased the reliance on automobiles. These have contributed to an increase in traffic congestion, outdoor air pollution, rising greenhouse gas emissions and poor public health. To make India's cities more sustainable and liveable in the future, administrators will have to find ways to shorten commuting distances and increase the use of public transport networks that is rapid and efficient to better urban productivity.

2. Safety and Accessibility

According to the World Health Organization (WHO), 10% of the world's road fatalities (130,000) occur in India alone (World Health Organization 2015). Traffic crashes occur every minute, and a life is lost every 3.7 min (Banzal et al. 2015). Such data means that the urban planners and decision makers should consider safety not only on the road but also in the surrounding environment. Accessibility and Safety are key components for ensuring the security of megacities.

3. Real Estate Developments

As more and more people move into cities of India, the requirement of safe places to live, commute and work will only increase. There's already a projected shortage of 18.78 million households in India between 2012 and 2017 (Tiwari et al. 2016). Real estate developers will try to fill this gap, which means that India's urban landscape will be massively influenced by their desires to make profits. So investment of private sectors in developments that provide access to sustainable transport and safer roads should only be encouraged, as it will benefit the state machinery setup a more secure urban sprawl.

Asia's cities will accommodate an additional 44 million people every year (Urbanization and Sustainability in Asia: Case Studies of Good Practice 2006). But the urban infrastructural needs in Asia are estimated to be over US$60 billion per year for water supply, sanitation, solid waste management, slum upgrading, urban roads and mass transit systems that are not met yet (Special Evaluation Study on Urban Sector Strategy and Operations 2006). Urban activities generate close to 80% of all carbon dioxide (CO_2) as well as significant amounts of other greenhouse gases which contribute to climate change (Report on the Dialogue on "Energy: Local Action, Global Impact" at the Third Session of the World Urban Forum. Direct sources of greenhouse gas emissions include energy generation, vehicles, industry and the burning of fossil fuels and biomass in households. Research shows that CO_2 emissions from the transport sector will triple in Asia over the next 25 years as the increase in the number of vehicles doubles every 5–7 years (Energy Efficiency and Climate Change Considerations for On-road Transport in Asia 2006). This increasing trend is unsustainable, and new measures to transport the public and mitigate the use of own vehicles should be encouraged. If we look at other environmental issues such as water and wastewater, solid waste management, slum management and air pollution control, we will come to a similar conclusion.

Picture 1 Rendered image of Meixi Lake eco-city in Hunan province (Image Courtesy: Kohn Pedersen Fox Associated)

There needs to be a change in thinking and, in some cases, even a radical shift in the way cities are managed should be incorporated. Only then can Asian cities be sustainable and liveable.

China in the race to modernise itself, urbanise and ascend economically has created entire districts of the country to be rendered as ecological wastelands. The air is deadly, the soil is toxic, the water is undrinkable, great lakes and rivers are disappearing, coastal wetlands have been decimated, and the cities themselves are becoming heat islands. Simply living in many of the China's cities is a health hazard. To end this rampant ravage on the environment, China is engaging in building new eco-cities, many of which are stand-alone, self-contained satellite developments outside the larger urban cores. This may seem counter-intuitive to the track record this country holds in the world's eco market, but China really has no choice. Over the coming decades, it has been estimated that 50% of China's new urban developments will be stamped with labels such as 'eco', 'green', 'low carbon' or 'smart' (Li and Yu 2011). One such city is the Meixi Lake eco-city in Hunan province. Picture 1 shows an urban planner's perspective of the Meixi Lake eco-city. Picture 2 shows the construction of the Meixi Lake eco-city.

The aim of this eco-city is to use renewable energy, urban agriculture, rainwater collection and a host of other sustainable technologies to cater to the city's needs. While all these measures sound good, the question to be asked is if the solution to build hundreds of these new cities actually an effective way of improving environmental conditions, as it involves clearing out massive swaths of farmland,

Picture 2 Construction of Meixi Lake eco-city (Image Courtesy: Wade Shepard)

demolishing rural villages and relocating thousands of nearly self-sufficient peasants. In their current form they are simply not bettering the environment as they are too small, too expensive and too remote. They may also have a political agenda, and some builders are going to benefit from such green wash philosophies. A better solution to creating new cities is to add a new dimension to existing cities and bring in technology, like seasonal energy storage and heat capture, rainwater collection, drinking water recycling or desalination, grey and black water systems, urban agriculture, sky gardens, distributed energy plants, waste energy recovery systems, thermal insulation, traffic-less downtowns, new modes or methods of public transportation, etc. Richard Brubaker quoted that 'Eco-cities should be the petri-dish by which all lessons for the megalopolises are learned and scaled'. 'If we're not learning anything and we're not scaling anything, then the eco-city is a distraction'.

4 The Smart City Philosophy

From Beijing's and Delhi's air pollution levels hitting hazardous levels, to Chennai's flood situation, to Jakarta's never ending traffic snarls, Asia's burgeoning megacities in 2016 have a huge problem to tackle. The answer is to be proactive rather than reactive when it comes to planning and building infrastructure. A counter to such situations is the idea of 'smart cities', which is echoing as the buzzword in some developing nations of Asia. With India and China planning to build hundreds of these cities, this philosophy is getting stronger and stronger by the day. While Chennai's drainage system, Jakarta's road infrastructure and Delhi's and Beijing's smog situation have thrown the spotlight on just how much work needs to be done, Asia is expected to take the lead in the construction of smart cities over the next century. With a predicted 62 megacities (cities with more than ten million people) by 2025, up from the current 23, Asia has no choice but to figure out the best way of using technology to help its citizens live better. But what is a smart city? Or what is the perception about smart cities in the minds of urban planners?

The term 'smart city' is quite broad and generic when it comes to its definition. Smart can mean many things in many situations and contexts, but most people agree that cities need to satisfy three things in order to be smart:

1. Using technology to drive or deliver services
2. Using technology to reduce costs for service that are being provided
3. Using technology to allow citizens to involve and improve the running of the city

A small island country like Singapore is probably the most famous example of a smart city in today's chaotic Asia surrounding it. The island country was named as the top global smart city in 2016 Smart City Asia Pacific Awards (SCAPA). Smart cities aim to engage with multiple functions like digital information and communication, multiple technologies and Internet of Things (IoT), etc. to manage a city better. This can refer to practically every aspect of the city, from local government departments to public services like schools, hospitals, transportation, waste management, law enforcement and security networks, etc. Singapore may have the advantage of geographic size, and with only over six million people and less than 1000 km^2, it provides a testbed to roll out new technologies and apply them differently and efficiently. It is also backed by a strong and committed government and auxiliary agencies, which support the cause of sustainable research and development. Singapore's smart nation project relies heavily on cloud computing in its infrastructure. According to the Info-communications Development Authority (IDA) of Singapore: 'Singapore's cloud adoption has grown from 24.6% in 2013 to 28.9% in 2015 with a strong adoption from both SMEs and enterprises according to a 2015 cloud adoption survey report by AMI Partners' (Ng 2016).

China's megacities may have problems of air pollution, chaotic traffic congestions and disparity in income distribution but some other problems like violent crime, slum proliferation, etc. that have been mitigated due to the construction of new cities. Some of the techniques used to avoid these problems are:

1. New Housing Constructions

Although the 'biggest bubble ever' jargon was pointed out by media houses when China began construction of new residential colonies, the housing market is actually quite stable than one may think. The ability of the country to provide modern accommodation for millions aspiring to become urban dwellers has prevented the proliferation of large shanty shacks and slums.

2. Upgraded Public Transport Network

If the public transport network does not support the infrastructure of the nation, it could spell doom for city planners. Examples of such chaos can be seen in the fast developing cities of Jakarta and Manila where automobile gridlock brings the city to a standstill. The ability to move efficiently through an urban space is paramount to generate more opportunity and work more efficiently. China's new cities are being developed on a central public transportation model that includes extensive underground subway networks and overhead feeder systems which ensure citizens efficient options to move around in lieu of using cars.

3. Land Use Zoning

Mumbai's floor area ratio (FAR) was around 1.33 throughout most parts of the city. That means that the city had to limit its construction to low-rise buildings some years ago. This led to the proliferation of overcrowded slums. Chinese cities in contrast allowed higher FAR in its urban dwellings, which meant that high-rise buildings left room for ample green space and less urban living congestions. China also implemented urban growth on newly annexed land outside of traditional urban cores, so the setting up of places like the new Pudong Area in Shanghai moved urban centres away from the existing city epicentres. Creating of many downtowns in an existing city is one of the solutions that should be followed to avoid congestions in a particular part of town. The suburban downtown must have a masterplan with extensive commercial, industrial, residential and technical centres.

4. Economic Incentives for Trade Zones

Chinese cities that are most business friendly offer attractive tax breaks as incentives and are likely to attract the bulk of domestic and foreign investments. Many cities in China are establishing special economic trade zones outside traditional urban core areas. The most successful story is of Shenzhen, which was started as a special economic zone (SEZ). This attracted many global brands to set up their manufacturing hubs on the city's periphery and provided employment to millions of Chinese residents.

Other improvements that can be carried out in existing cities to make them smart are by developing strategies to decongest existing roads. While limited funds and space are always a huge challenge, human health, pollution and connectivity to drive businesses should take priority. CCTVs, traffic signals, flow of traffic and other data sources, etc. should be combined to give the city administration a better idea of where the arterial choke points are. Traffic speed and volumes should be analysed in real time to get a picture of what the traffic looks like at any point of time in the city. IBM is currently working on making this more accessible, and Singapore's Smart Mobility 2030 is probably the best example of a city trying to incorporate these inputs and build a smarter transport network.

Songdo, in South Korea, is a smart city that is being built from the ground up. It is grand in ambition and in scale. Many countries are closely watching this city's development as the city features some of the cutting edge technologies in urban

Picture 3 A view of Songdo's Central Park in South Korea (Image Courtesy: Dongho Kim)

planning and development. The city is being built around an airport, to reduce travel time. It takes 15 min to Seoul's Incheon airport, closer than Seoul itself. Built-in sensors on streets and buildings are provided to monitor and manage human traffic. Up to 40% of the city's space will be green space. There are pipes connecting offices and apartments which will manage and recycle waste. The city itself is expected to be finished by 2018 but at a cost of an estimated US$35 billion. The 10-year project is expected to be one of the most grand and expensive development projects ever undertaken. An image of the Songdo Central Park is shown in Picture 3.

The first Zero Carbon City named Masdar City project is also being built close to Abu Dhabi in the United Arab Emirates (UAE). Masdar City is seen as the flagship project for Abu Dhabi to move towards a more sustainable and less oil-dependent future. Some of the sustainable design elements that are incorporated in the Masdar City are a 45-m wind tower, based on traditional Arabian wind tower designs that helps keep the street levels 15–20 °C cooler than surrounding areas. Solar panels on the roofs power most of the city. Making most of the city car-free by providing driverless electric pods that transport users back and forth to destinations and using locally available building material to construct the city are some of the sustainable features the city boasts of. This smart city too is being constructed at a whooping cost of US$ 16 billion and is expected to be completed by 2025 (Picture 4).

Songdo and Masdar are two examples of governments trying to build a smart city from the scratch and at an astronomical cost. These surely don't give replicable answers to cities with existing problems. Even if both cities are fully built and functional, they will be home to less than a million people each; and Asia's problem

Picture 4 Driverless electric cars in Masdar City (Image Courtesy: Michael Bumann)

today are cities with more than ten million people. Driverless electric pods cannot help Jakarta ease its clogged roads. They cannot give better solutions to Chennai to manage its rainwater nor can solar energy in its present form help Beijing create green space and manage air pollution. Cities like Songdo and Masdar City are more like testbeds; the actual practical model could be developed taking some cues from such developments. Asian cities are still a long way from reaching their utopian goal, but practical and doable solutions are needed to solve and conquer the long battle of getting Asians to have better living conditions.

5 The India Story of Smart Cities

'Smart city' has emerged as a buzzword in India during the last 3 years ever since Prime Minister Narendra Modi outlined his vision for creating 100 smart cities in the next decade. The smart cities that are coming up in India are not new cities that are being built from the scratch; they are existing towns, smaller cities and a few megacities that are being infused with technology-driven operational models. The aim is to change the way a city functions by providing smarter solutions to the existing traditional approaches of performing tasks. This is being done to improve standards and provide basic sanitation, clean drinking water, effective and efficient waste disposal mechanisms, effective public transport systems, uninterrupted power supply, etc. According to a United Nations research, India will add another 404 million people by 2050 (United Nations 2014a). That is equivalent to adding two populations of Singapore every year until 2050. This rate of urbanisation will have large repercussions on existing cities, and attending to the needs of education, healthcare and security will become a challenge. Developing smart solutions to use limited resources is thus a priority. Redevelopment and reshaping of existing towns and planning of smaller city developments are the need of the hour for India's smart city programme. Empowering existing urban settlements with technological solutions to problems is the success formula for India's megacity problems. But the success of the smart city mission can only be achieved when meticulous planning meets technology-inspired solutions. Using of sensors to create smart grids and information from data analytics will allow the city infrastructure and services to meet the population's requirements and citizen's demands efficiently and reliably. The Internet of Everything (IoE) is projected to be a US$1.5 trillion business globally (Bradley et al. 2013). So the smart cities initiative should promote the idea of cities becoming R&D centres for hardware, software and urban services. They should also promote the design, development and manufacturing of locally usable products in the country, thereby creating a market for exports. Street lighting in India takes a back seat even in its megacities, where major stretches of roads are not lit. This causes an increase in crime and is a nightmare for security establishments. According to McKinsey, street lighting accounts to only 1.5% of India's total electricity consumption (Baumgartner et al. 2012). While the number is a good indicator of low-energy consumption, the absence of street lighting in major parts of megacities is not welcome. Cities that can use networked motion detection systems can save 70–80% of lighting electricity (Bradley et al. 2013). Smart street lighting initiatives can also reduce crime in the area by 7% because of better visibility and more content citizenry (Bradley et al. 2013). The energy consumption by Indian buildings is 40% of the nation's annual consumption, and by 2030 this figure is expected to rise to 50% (United Nations 2014a). With an estimated 700–900 million m^2 of new residential and commercial space needed to satisfy this population, the energy footprint will only increase with time. A technological improvement that can be suggested for improving traffic problems in India can be by using embedded and networked sensors for parking. This technology can relay real-time information to drivers and direct them to available parking spots. All downtown areas in megacities have more than half its cars circling around looking for parking options. Such a technological solution can help reduce pollution, congestion and fuel consumption. Towns can also become a part of a regional cluster and offer remote management services like infrastructure monitoring to other surrounding megacities. These towns can eventually develop into becoming centres of excellence for particular domain skills.

Smart cities will also have to believe in the model of public-private partnership (PPP) and build-operate-transfer (BOT) to build and operate infrastructure. Effective PPPs and BOTs have been seen in the construction of expressways and other transport infrastructure. The Mumbai-Pune Expressway and Yamuna Expressway connecting the nation capital Delhi to the city of The Taj Mahal in Agra are some such successful models. The advantage of having such deals is that the private sector's risk-taking capacity and access to funding will increase while ensuring that economics of the state will serve the public good. The concept of foreign direct investment (FDI) by the Indian government is also a welcome move. This gives foreign skill and expertise an opportunity to invest in India and reap the benefits monetarily. The country in turn will acquire good and state-of-the-art facilities. In a technology-intensive future, job creation and GDP growth will depend on the steady incubation of new and innovative companies that can scale up to become global game-changers. Smart cities should encourage start-ups that focus on solving urban challenges like traffic, crime, energy conservation, etc. by leveraging technology. India must also create technologies that will help government bodies secure revenues, eliminate overlapping roles and explore investment opportunities. Parallel administrative power centres like government functionaries and mayors should have clear-cut jurisdictions of work and administrative authority. Accountability and responsibility should be introduced in a bid to change existing patterns, and the security of government jobs must be leveraged for this cause. On the contrary, administrators and planners who do well and contribute to successfully implementable models must be rewarded monetarily to encourage others to look upto and get motivated. The ongoing Delhi-Mumbai Infrastructure Corridor (DMIC) initiative, where the government has sought to develop seven smart cities across six states, is making progress because it is operated like an enterprise where the project is being overseen by a CEO. Managing a city's physical infrastructure should be done by setting new standards for street lighting, parking, garbage bins and collections, etc. The same should be monitored on performance and requirements; with regular and enforced maintenance being carried out. City authorities must use their political will to mandate via regulations the use of smart technologies while safeguarding citizens and contentious issues around liability, security and privacy.

Indian cities today accommodate nearly 31% of the country's current population and contribute to 63% of its GDP (Registrar General 2011). Urban areas are expected to house 40% of India's population and contribute to 75% of GDP by 2030 (Registrar General 2011). This would require a comprehensive development of physical, institutional, social and economic infrastructure in order to improve the quality of life and attract people and investments. Development of the smart cities in India is a step in that direction. The focus on developing smart cities is to address most pressing needs of society and improve lives by increasing opportunity in digital and information technology, urban planning, public-private partnerships and policy changes. The Smart Cities Mission is driven with the objective of promoting cities' core infrastructure thereby increasing the quality of life of its citizens and providing a clean and sustainable environment. This is an innovative and new initiative by the Government of India to drive economic growth and enable local developments. It aims to bring adequate water supply and assured electricity supply; improve sanitation and solid waste management; make urban mobility and public transport efficient; provide affordable housing for the poor; make robust IT connectivity and digitalisation; improve and make good governance, especially e-Governance and citizen participation; deliver a sustainable environment to its citizens; provide good health and education; and provide safety and security to citizens, particularly women, children and the elderly. A total number of 100 smart cities have been identified and distributed among the states and union territories (UTs) across the country on the basis of an equitable criterion. The formula gives equal weightage

(50:50) to urban population of the state/UT and the number of statutory towns in the state/UT. Based on this formula, each state/UT will, therefore, have a certain number of potential smart cities, with each state/UT having at least one. The number of potential smart cities from each state/UT will be capped at the indicated number. This distribution formula has also been used for allocation of funds under Atal Mission for Rejuvenation and Urban Transformation (AMRUT). The distribution of smart cities will be reviewed after 2 years of the implementation of the Mission. Based on an assessment of the performance of states, some reallocation of the remaining potential smart cities among states could be done by the Ministry of Urban Development. An average of Rs. 100 crore (US$15 million) per city per year is being given for 5 years in order to develop such technological resources. The financial support extended in total would be around Rs. 48,000 crore (US$7.2 billion) for the smart city mission. An equal amount, on a matching basis, will have to be contributed by the state/UTs. Therefore, nearly US$14.4 billion of funds will be available for smart city development projects. There have been three rounds already, where cities have been picked for funding. They are as below and can be read as <city> (state they belong to):

First Round:

Bhubaneswar (Odisha); Pune (Maharashtra); Jaipur (Rajasthan); Surat (Gujarat); Kochi (Kerala); Ahmedabad (Gujarat); Jabalpur (Madhya Pradesh); Vishakhapatnam (Andhra Pradesh); Solapur (Maharashtra); Davangere (Karnataka); Indore (Madhya Pradesh); New Delhi (New Delhi); Coimbatore (Tamil Nadu); Kakinada (Andhra Pradesh); Belagavi (Karnataka); Udaipur (Rajasthan); Guwahati (Assam); Chennai (Tamil Nadu); Ludhiana (Punjab); and Bhopal (Madhya Pradesh)

Second Round

Lucknow (Uttar Pradesh); Warangal (Telangana); Dharamsala (Himachal Pradesh); Chandigarh (Chandigarh); Raipur (Chhattisgarh); New Town, Kolkata (West Bengal); Bhagalpur (Bihar); Panaji (Goa); Port Blair (Andaman and Nicobar Islands); Imphal (Manipur); Ranchi (Jharkhand); Agartala (Tripura); and Faridabad (Haryana)

Third Round:

Amritsar (Punjab); Kalyan (Maharashtra); Ujjain (Madhya Pradesh); Tirupati (Andhra Pradesh); Nagpur (Maharashtra); Mangaluru (Karnataka); Vellore (Tamil Nadu); Thane (Maharashtra); Gwalior (Maharashtra); Agra (Uttar Pradesh); Nashik (Maharashtra); Rourkela (Odisha); Kanpur (Uttar Pradesh); Madurai (Tamil Nadu); Tumakuru (Karnataka); Kota (Rajasthan); Thanjavur (Tamil Nadu); Namchi (Sikkim); Jalandhar (Punjab); Shivamogga (Karnataka); Salem (Tamil Nadu); Ajmer (Rajasthan); Varanasi (Uttar Pradesh); Kohima (Nagaland); Hubli-Dharwad (Karnataka); Aurangabad (Maharashtra); and Vadodara (Gujarat)

There is tremendous potential in India to build an effective urban ecosystem and enable our burgeoning urban areas to become smart by using digital technology. This in turn will create employment opportunities and contribute to economic growth through innovation. Our cities are fast becoming the defining units of human habitation. How smartly we build, manage and operate our cities will be the single biggest challenge of this century and will determine our people's future. We owe it to our future generations to make our cities smart through the use of technology.

References

Banzal, R.K., Jaiin, A., Yadav, J., Dubey, B.P.: Pattern and distribution of head injuries in fatal road traffic accidents in Bhopal region of Central India. J. Indian Acad. Forensic Med. **37**(3), 242–245 (2015)

Baumgartner, T., Wunderlich, F., Jaunich, A., Sato, T., Bundy, G., Grießmann, N., Kowalski, J., Burghardt, S., Hanebrink, J.: Lighting the Way: Perspectives on the Global Lighting Market (2012)

Bradley, J., Reberger, C., Dixit, A., Gupta, V.: Internet of everything: a 4.6 trillion public-sector opportunity. Cisco White Paper (2013)

Energy Efficiency and Climate Change Considerations for On-road Transport in Asia (Philippines: Asian Development Bank.: Retrieved from http://www.adb.org/Documents/Reports/Energy-Efficiency-Transport/default.asp (2006)

Li, H., Yu, L.: Chinese eco-city indictor construction. Urban Stud. **7**, 015 (2011)

Ng, K.: What is a Smart City and how does Asia rank among the world's best. TechWire Asia. Retrieved from http://techwireasia.com/2016/08/smart-city-asia/ (2016)

Poverty, E: Millennium development goals. United Nations. Available online: http://www.un.org/millenniumgoals/ (2015).

Registrar General, I: Census of India 2011: provisional population totals-India data sheet. Office of the Registrar General Census Commissioner, India. Indian Census Bureau (2011)

Report on the Dialogue on "Energy: Local Action, Global Impact" at the Third Session of the World Urban Forum, Vancouver, Canada, 22 June.: Retrieved from http://www.unhabitat.org/cdrom/dialogues/3b_r.html (2006)

Sankhe, S., Vittal, I.: India's Urban Awakening: Building inclusive Cities. Sustaining Economic Growth, McKinsey Global Institute Report (2010)

Special Evaluation Study on Urban Sector Strategy and Operations (Philippines, Asian Development Bank): Retrieved from http://www.adb.org/Documents/SES/REG/sst-reg-2006-03/ses-usso.asp (2006)

Tiwari, P., Rao, J., Day, J.: Housing development in a developing India. In: Development Paradigms for Urban Housing in BRICS Countries, pp. 83–139. Palgrave Macmillan UK (2016)

United Nations, Department of Economic and Social Affairs, Population Division: World Urbanization Prospects: The 2014 Revision, Highlights (ST/ESA/SER.A/352) (2014a)

United Nations, Department of Economic and Social Affairs, Population Division: Population facts: Our Urbanizing World: August 2014 (No. 2014/3) (2014b)

Urbanization and Sustainability in Asia: Case Studies of Good Practice (Philippines: Asian Development Bank): Retrieved from http://www.adb.org/Documents/Books/Urbanization-Sustainability/default.asp (2006)

World Health Organization: Global Status Report on Road Safety 2015. 2015. Geneva (2015)

Xueqin, J.: China's Mega City Problem. The Diplomat. Retrieved from http://thediplomat.com (2012)

Development of Smart Cities from Fiction to Reality in Member States of the Gulf Cooperation Council

Jihad Awad, Afaq Hyder, and Adi Irfan

Abstract This study analyzes and outlines the fundamental components of a smart city where information and communication technology (ICT) is synchronized with traditional infrastructures by utilizing new technologies. This study developed the criteria of ICT advancement that reformed the functions and management of a city. These criteria defined opportunities that facilitated rapid explicatory interaction among common citizens, governments, businesses, and several agencies. This study found that ICT technologies facilitate essential engagement in the design and planning of a modern city. This study examined the possible conceptual basis for establishing smart cities. This study also established the viable relation between a fictitious futuristic city and a smart city. Several models of smart cities were reviewed, compared, and summarized. This study highlighted the effectiveness of new technologies in addressing urban challenges, such as urban governance and organization, transportation, energy, and revenue collection. The first part of this study presented the definition and general understanding of smart cities, explained the science of smart cities, reviewed the conceptual basis of these cities, and focused on future ideas about effective digital networks to manage a modern city through smart technology.

1 Introduction

The concept of smart city as fiction or reality has been debated in the past. The answer to such a debate is clear because we are already living and experiencing the technological fractions for smart cities. For example, daily smart activities could be experienced when crossing a tollgate with a radio-frequency credit chip fixed

J. Awad (✉) • A. Hyder
Department of Architecture, Ajman University, Ajman, UAE
e-mail: dr_jihadaa@yahoo.com

A. Irfan
Department of Architecture, National University of Malaysia, Bangi, Malaysia

S.T. Rassia, P.M. Pardalos (eds.), *Smart City Networks*, Springer Optimization and Its Applications 125, DOI 10.1007/978-3-319-61313-0_4

on a windscreen, paying service bills through cellular phones, mobile banking, controlling internal climate through Wi-Fi systems, printing without utilizing cables, and watching television without using conventional broadcasting cables. Smart technology is used when constructing smart grids with smart meters, using renewable sources of energy, utilizing waste to create renewable sources of energy, and adopting sensor-controlled light and water supply fixtures. Humans use smart infrastructures, such as smart roads and modes of transportation with integrated sensors and CCTV for intelligent traffic management systems, satellite-controlled traffic warnings, and navigation systems.

Future smart cities will be considered hubs of total smart technology and systems applied in a certain area. These cities will soon emerge as neighborhoods where residents control primary and secondary routines of daily life by utilizing ICT technology. However, this development will replace personal appearance with hologram technology.

Smart city is an ICT tool that can be used by governments or private sectors to improve the quality of life by increasing the efficiency of city management, upgrading the allocation of resources, and promoting economic development, sustainability, and innovation. The growing world population, particularly in South Asia, demands drastic changes in conventional paper-based management system. According to the United Nations (UN) (2015), around 70% of the world's population will be concentrated in urban centers by 2050.

Aside from South Asia, the member states of Gulf Cooperation Council (GCC) are also expected to achieve one of the highest urbanization rates in the world. Luxurious lifestyle, peace, and tax-free environment lure people, particularly those from South Asia, to migrate to GCC countries. This phenomenon resulted in issues, such as increased number of cars on roads, traffic congestion, immigration influx, pollution, resource management, and resource depletion. GCC regions also experienced inflation in essential supplies, such as power and water, because governments downsized subsidies on certain local products, particularly fuel. This phenomenon caused the prices of other essential commodities that rely on transportation and fuel to increase. GCC countries believe that these issues could be resolved through smart systems and renewable sources of energy. The reduction of fuel prices affected petroleum-based economies in GCC, but the various smart projects launched in these regions show that GCC governments are committed to switch from traditional management to smart management technology and systems.

The member states of GCC aim to adopt effective ways to conserve their natural resources and manage existing cities by utilizing alternative sources of energy and smart measures. Governments are keen on developing new clusters of living as smart cities and establish these cities as a hallmark for future development in these regions. GCC member states, such as the United Arab Emirates (UAE), Kingdom of Saudi Arabia (KSA), Kuwait, and Qatar, initiated projects to establish high-tech cities. These states recognize that smart cities are necessary for survival and they quickly established infrastructures that are adaptive to sudden changes in the supply–demand cycle.

2 Conceptual Basis of a Smart City

Smart city is a novel buzzword that may refer to a neighborhood or an urban development infrastructure that integrates multiple information and communication technology (ICT) solutions to manage city resources securely. Mobile phone technology seemed fictitious in 1985. However, the advent of smartphones and other wearable technology at present exceeded works of fiction. The term "smart" was used to describe beautiful, fast, spontaneous, and reliable individuals and not devices or lifestyles. However, the world drastically changed. People now expect to experience fast, safe, and streamlined services while working, learning, playing, purchasing, traveling, managing finances, maintaining fitness, or watching television. Therefore, "smart" is used to describe fast, alert, safe, and responsive technology for daily use in modern age.

The perception of smart city transports the reader to the era of early artistic age of science fiction where artists conceptualize future in another planet. These imaginative illustrations show that the conception of artists was close to the structures that were later built and launched in contemporary world.

However, not all of the advancements imagined by science fiction authors and artists have been achieved. Buildings and different urban forms similar to the illustrations of science fiction artists are built. These artists presented conceptual building designs for living on other planets, but settlements in planets (or in space) are inexistent. However, the modern concept of smart cities flatters the concepts of science fiction developed in the early twentieth century, as shown in Figs. 1 and 2.

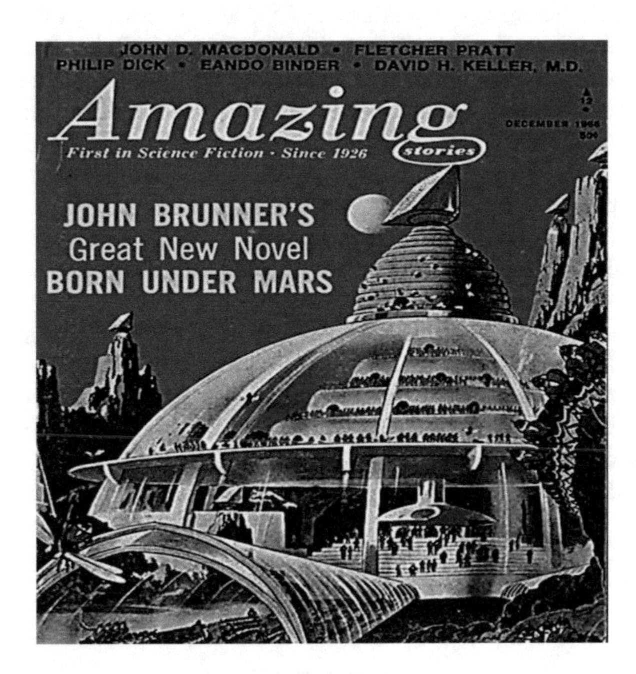

Fig. 1 Science fiction concept (Source: Raitt et al. 2004)

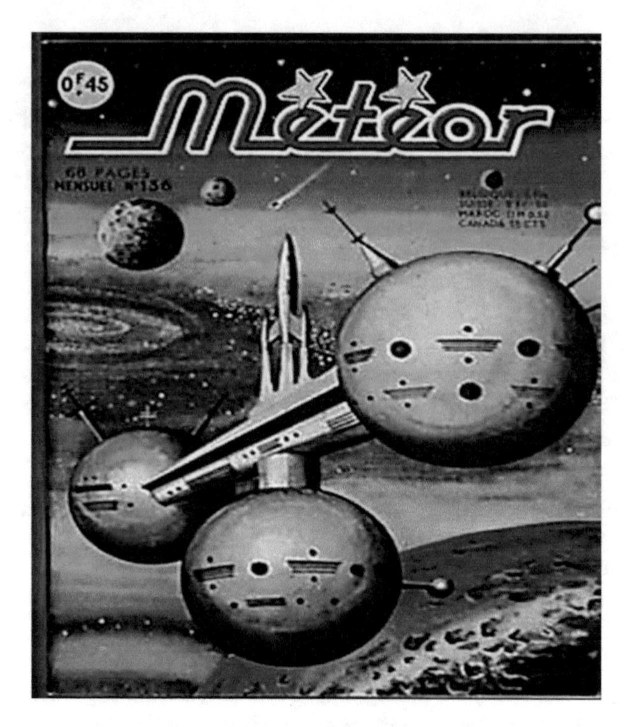

Fig. 2 Science fiction concept (Source: Raitt et al. 2004)

Early science fiction authors, artists, and illustrators described the concepts of future cities based on the limited scientific knowledge available at the time. The works of previous generations of science fiction artists and authors were enhanced by present-day designers and planners. Modern designers generally portray similar concepts of science fiction in the early twentieth century. The imagination of science fiction artists and authors created a niche for modern-day scientists and technologist to develop similar environments and practical gadgets for smart cities.

Caplescu (2012) describes science fiction as a futuristic way of thinking, which is symbolized by future advancements in technology and a future society developed from current paradigms. The future is usually portrayed as a statement of present-day issues regarding society, environment, politics, economics, and religion or questions about progress in various fields of science. Speculative technology in films inspired researchers to invent devices. For example, the first mobile phone, which was developed by Motorola, was actually a flip phone that may have been inspired from the communicator device in *Star Trek* (the original 1966 TV series), as shown in Figs. 3 and 4.

Fig. 3 Communicator in Star Trek (http://www.eyeonstarwars.com/episode1/picture/e196.jpg. Source: Startrek 1966)

Fig. 4 Motorola first flip phone in 1989) (http://fm.cnbc.com/applications/cnbc.com/resources/img/editorial/2015/07/07/102815067-Untitled1.530x298.jpg?v=1436295299. Source: MicroTAC 1989)

However, this phenomenon could be viewed in an opposite manner because modern researchers believe that classical buildings and architecture served as inspiration for futuristic designs. According to Hanson (2005), the Art Nouveau movement, particularly the work of Gaudi, was used in designing Gungan City (Star Wars), as shown in Figs. 5 and 6. The Marin County Civic Center of Frank Lloyd Wright served as inspiration for the palace of Queen Amidala (Star Wars), as illustrated in Figs. 7 and 8.

Fig. 5 Gungan City (Star Wars theme) (http://nausheenhusain.files.wordpress.com/2012/03/5362241572_6b17f4b682.jpg. Source: Hanson 2005a)

Fig. 6 Casa Batllo in Barcelona, Spain, by Gaudi Italy (https://www.bluffton.edu/homepages/facstaff/sullivanm/spain/barcelona/gaudibatllo/windows.jpg. Source: Hanson 2005)

This discussion indicates that the basis of new smart city concepts and other futuristic designs could be traced to the ideas of science fiction artists and authors in the past. The concept of modern smart cities is limited to the ideas of top designers, but these conceptualizations could have also evolved from sketches of science fiction artists.

Fig. 7 Palace of Queen Amidala (Star Wars movie) (http://realestategals.com/wp-content/uploads/2015/12/StarWarsEpisodeI-OtohGunga.jpg. Source: Hanson 2005)

Fig. 8 Marin County Civic Center (From Frank Lloyd Wright. Source: Hanson 2005)

3 Global Smart City Programs

Modern-day planners and designers consider smart cities as impressively designed and viable solutions to urban problems, such as overpopulation and congestion, housing, time and resource conservation, and efficient management. Among these problems, rapid population growth in urban centers is the most significant concern that drives various industries, professionals, technologists, and designers to collaborate and seek solutions in smart cities.

Zhao (2016) predicts that 70% of the world population will live in cities by 2050; thus, sustainable urbanization became a policy priority for administrations across the world. ICTs play a crucial role in increasing efficiency across industrial sectors and enabling innovations, such as intelligent transport systems and "smart" water, energy, and waste management. The benefits of establishing "smart" technologies in an existing city or developing smart and sustainable cities from the ground up have been widely recognized. Zhao (2016) highlighted the issue of rapid urban growth and established a correlation between emerging challenges and the increasing pollution in urban centers by 2050. This study stressed the need to address urban challenges by adopting smart technologies and suggested the need for a global perspective in addressing this issue. Zhao (2016) recommended uniform policies or guidelines for the adoption of smart technologies to transform existing cities or develop new smart cities around the world.

Several countries developed models to adopt the features of smart cities on existing and new cities. These schemes are controlled by regional conditions and future challenges of a particular region. However, these models generally assume that smart cities should act as a hub with highly efficient mechanisms. These models also assume that smart cities are built on the horizontal and vertical integration of city processes by maximizing the data generated by IT systems.

3.1 Smart City Opportunities for the UK

The Department for Business, Innovation and Skills (BIS 2013) published a research and presented a prospectus for developing smart cities in the UK. BIS developed the following chart to explain the effectiveness of the smart city model in various sectors in the UK, as shown in Fig. 9. In addition, Table 1 shows the summary of model.

3.2 Smart City Model of the European Commission

The European Commission (2014) proposed a model to develop or promote smart cities around Europe. This model stressed the need to explore new sectors of smart development and highlighted the key sectors in which the smart technology would be introduced. The objective of this model, which is shown in Fig. 10, is summarized as follows:

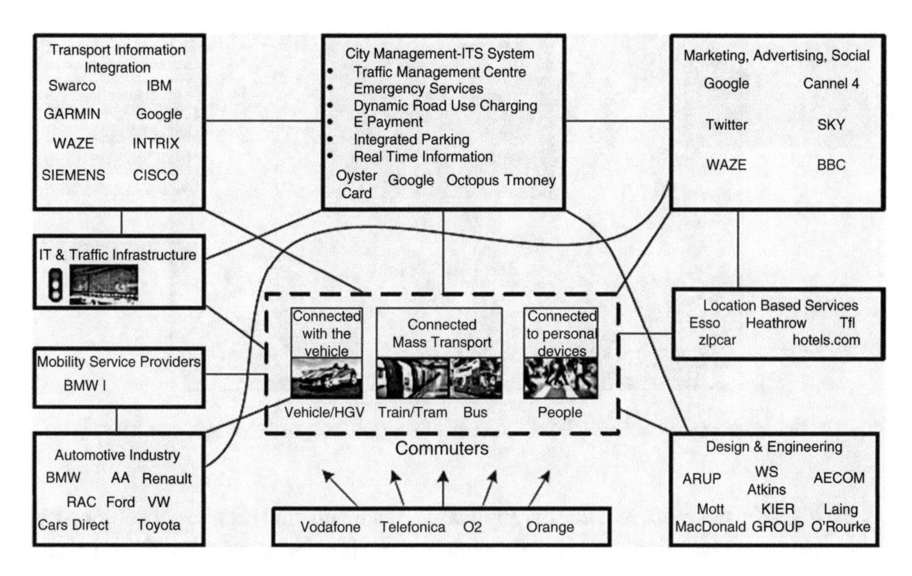

Fig. 9 Smart city management model in the UK (Source: BIS 2013)

Table 1 Smart city management model in the UK

City management	City administrators can reduce or control the increasing pressure and congestion of commuters and businesses through smart city applications. Smart management will improve the quality of life of individuals living and working in cities
Transport information integration	Smart traffic control will provide solutions on traffic management in cities. The smart transport mechanism will eventually optimize transportation, improve safety, and reduce congestion
Mobility service providers	This smart initiative will deliver customized transport solutions to city inhabitants. These measures are minor but are a growing player in industries that offer services to commuters
Automotive industry	This smart initiative is related with the automotive industry, which includes manufacturing, insurance, repair, and recovery. This initiative is interested in providing commuters with comprehensive services. Smart services and products aim to provide sustainable, green, and safe transportation
Design and engineering	This smart measure is involved in developing transport strategies, policies, and compliance. This sector aims to design and build transportation systems, which are optimally designed, incorporate best practices to reduce congestion, improve the quality of life, lower emission, and maximize mass transport
Location-based services	This smart measure is crucial. Urban dwellers will most frequently experience these types of services. This sector will provide services, such as beds, food, fuel, and transportation, to commuters. Commuters will be informed about available services before, during, or after their commute through smart technology
Marketing, advertising, social	Advertising companies will use smart technologies to influence customers to sell or provide a service

Fig. 10 European commission model of smart city (Source: European Commission 2014)

1. To identify, develop, and deploy replicable, balanced, and integrated solutions in energy, transportation, and ICT through partnerships between municipalities and industries
2. To encourage individuals and city administrators to utilize smart technologies, such as lighthouse projects, which will primarily target the large-scale demonstration of replicable smart city concepts in the city context where existing technologies or similar market technologies will be integrated innovatively
3. To promote near-zero or low-energy districts, integrated infrastructures, and sustainable urban mobility for communities

3.3 Smart Urban Solutions of the United Nations Economic Commission for Europe (UNECE) for Transitioning and Developing Countries

According to the United Nations Economic Commission for Europe (UNECE) (2015), high urbanization presents a serious challenge for the sustainable development of cities. Urban areas are responsible for a considerable amount of the world's energy consumption and net greenhouse gas emissions. Urbanization leads to the continuous expansion of urban areas and urban sprawl, which limit land for other uses. According to UNECE, these problems could be controlled by the smart city schema. Two models of smart cities were presented by UNECE. The first model describes the global approach of smart cities, which is followed by developed countries. The second model was proposed for developing economies. The key factors of both models are summarized in Tables 2 and 3.

Table 2 Key factors of global smart city model

Key factors	Rationale
Urban scape	Improving the urban environment (buildings, transportation, water, waste, and energy services by utilizing information and communication technology [ICT])
Good governance	Establishing the role of ICT in city governance (inter-sectorial cooperation; cooperation between national, regional, local authorities and other stakeholders; establishment of multi-stakeholder platforms); support networking with other cities
ICT-based social capital	Developing an ICT base for education and social and gender equality
Economic conditions	Promoting business-friendly policies and (ICT) infrastructure for poverty reduction and employment generation
Tourism	Encouraging urban tourism through ICTs

Table 3 UNECE smart city model for developing economies

Key factors	Rationale
Controlling the city challenges	Lessening the vulnerability of cities
Cutting-edge technologies	Latest and quick-win initiatives that are easy to implement will be introduced to efficiently utilize resources at the disposal of municipalities
Public and private partnership	Cooperation between the public and private sectors in research will be improved, and innovative solutions will be implemented to achieve sustainable urban environments
City welfare	The well-being of city inhabitants will be enhanced, and the environmental quality of cities will be improved
Boosting the private sector	New markets, multiple projects, and financing opportunities across regions can be identified. The possible establishment of new industries and fields of business in developing transition countries will lead to economic growth, create working places, and increase the well-being of inhabitants
Replication of successful models	Reproducible examples will be created for cities with similar backgrounds and financing possibilities
Establishing the setup of professional expertise	A network of experts and city administrations across regions will be developed
Establishing the setup of institutions	A network of institutions at the regional, national, and local levels will be developed

3.4 UNECE Smart City Model for Developing Economies

Table 3 summarizes the smart city model for developing economies proposed by UNECE. UNECE emphasizes that the effects of adopting this model on developing cities are long term, and smart solutions will be extended to low- and medium-income transition economies.

3.5 Smart City Model of the USA

The official web pages of the USA offer limited information on smart city models. The following table presents the smart city model of the USA based on various websites (Table 4).

Several smart city models were discussed in this section. Most of these models address the challenges of the growing population. Most of the factors mentioned in these models aimed to address urban congestion and reduce the load of city administrators. The majority of the models emphasize the need to use ICTs to resolve current and future urban issues. However, UNECE, which is a department under the UN, proposed two types of smart city models. The first model is currently implemented and the most adopted model in the developed world. The second model outlines the concept of smart city for the developing world. Unlike the first model, UNECE suggests that the smart city model should boost the private sector, replicate successful models, and establish a system of professional expertise. These recommendations suggest that the UN views smart cities in the context of ICT usage and aims to foster economic development and population welfare through this model. The smart city model of the USA differs from other models because it appears comprehensive and effective in conveying long-term benefits to citizens. For example, factors related to healthcare and disaster recovery are crucial in urban life because aging and climate change are significant challenges in the contemporary world.

Table 4 US model of smart city

Key factors	Rationale
Home/building	Convergence of smart home and building architecture SCALE – safe community alert network smart home/business gateway platform smart power, smart light – made in Detroit smart rooftops service enablement provider
Climate/ environment	Enhanced water distribution infrastructure smart cities USA
Disaster recovery	Event management for smart cities SERS – smart emergency response system smartphone disaster mode
Manufacturing	Smart manufacturing smart shape technology
Transportation	Applied robotics for installation and base operations smart roads smart vehicle communications Southeast Michigan smart transportation
Healthcare	Closed loop healthcare connecting smart systems to optimize emergency neurological life support project boundary SCALE – safe community alert network
Security	Cyber-secure SyncroPhasers with security fabric smart energy CPS the agile fractal grid
Energy	Cyber-secure SyncroPhaser with security fabric smart energy CPS smart power, smart light – made in Detroit smart rooftops transactive energy management

Source: Improved from Smart America (2014)

4 Smart Cities in GCC

In the past two decades, GCC, particularly UAE, underwent rapid urban growth development and became the global hub of real-estate investment. During this period, GCC established colossal buildings and corporate businesses. However, rising fuel prices pushed oil-rich GCC member states to identify alternative economic resources to maintain the economic performance of the region. The oil-rich GCCs aim to transform their petroleum-based economy to other forms of economy, such as tourism, economic development, real estate, and manufacturing. GCC countries believe that smart cities can support this transformation and create a knowledge economy. GCC countries expect that this transformation will provide the young population with the skills required for professional and senior-level jobs.

According to the study of International Data Corporation (2015), GCC countries aim to become "smart" to achieve goals or realize a vision, such as Plan 2021 of Dubai, Vision 2030 of Dubai, and the efforts of Saudi Arabia toward economic diversification by establishing economic cities. The study identified the interesting facts and plans of three member states of GCC, i.e., UAE, Saudi Arabia, and Qatar. Table 5 highlights the key initiatives of these countries.

4.1 *Smart City Model of UAE*

Similar to other countries, UAE also prepares to address challenges related to rapid urbanization and the monetary system. The UAE recognizes that the provision of smart and user-friendly ICT services in cities will help manage cities using less

Table 5 Smart initiatives of GCC

Country	Initiative	Narrative
UAE	Strategy for Dubai's smart	Transformation – six pillars covering 100 initiatives in transportation, communication, infrastructure, electricity, economic services, and urban planning
Saudi Arabia	Economic and smart development	The kingdom is constructing a railway network that extends to nearly 663 km
		Modernizing older cities, such as Jeddah, Riyadh, Makkah, and Medina; enabling cities in utilizing GIS and traffic management systems for urban planning, traffic management, and overall safety
		The Kingdom also established its first smart city, namely, Yanbu, which will accommodate 300,000 residents and create as many as 80,000 jobs
Qatar	Smart development	Smart city investments include prominent projects, such as Lusail City, Msheireb, Barwa City, Energy City Qatar, and Pearl Qatar

Source: Improved from IDC (2015)

resources and will raise the quality of living of city dwellers. Technology will play a major role in achieving the smart and efficient cities. UAE adopts a unique approach to transforming its current cities to smart cities. The emirate of Dubai is leading the smart city movement in UAE. The smart city initiatives of UAE emphasized several factors, such as communication, total integration, sustainability, and cooperation. UAE believes that a united approach will succeed and help establish the country as a leading nation with many smart cities. Among the seven emirates of UAE, the emirate of Dubai is considered the technological, social, and tourist hub. Dubai initiates and sets a standard for the other emirates of the UAE. Dubai pioneers in setting the standards and factors that are suitable for smart cities in UAE. The government of Dubai is eager to declare the city as the first fully developed smart city in GCC. UAE can replicate the successful smart city model of Dubai to other emirates.

Researchers believe that Dubai is a regional GCC risk taker and a trendsetter in terms of adopting digital technology. Dubai adopted cutting-edge approaches in digital governance, utilized ICTs for development, adjusted policies and regulations to adapt to rapid societal changes and technological advancements, provided enabling infrastructures for Internet businesses, and created a hub for a knowledge economy that extends to the region (Salem 2016).

Dasani et al. (2015) quote His Highness Sheikh Mohammed bin Rashid Al Maktoum, Prime minister of UAE and ruler of Dubai: “Our goal is for the entire city’s services and facilities to be available on smartphones. We want to provide a better quality of life for all.” The smart concept of Dubai is two sided. One side depicts the technological aspect, whereas the other side portrays humanistic considerations. Dubai developed an interesting correlation between happiness and smart cities. The logo in Fig. 11 shows that the Dubai government believes that Smart Dubai (technology) will promote happiness (humanistic) among its citizens.

Fig. 11 Smart Dubai (Source: Salem 2016)

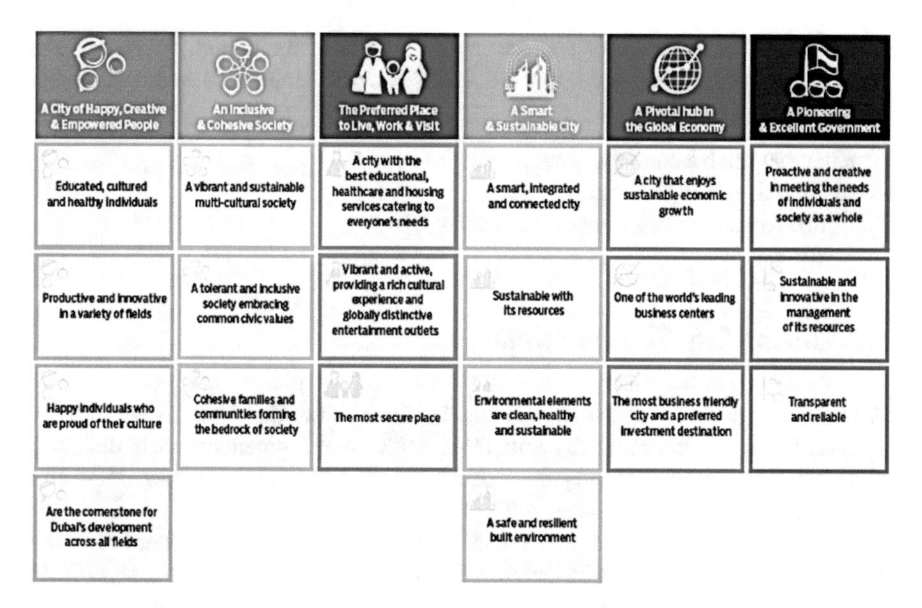

Fig. 12 Dubai (UAE) 2021 Plan (Source: Salem 2016)

Based on the philosophy of happiness through technology, the Dubai government proposed a plan for its smart city model, as shown in Fig. 12.

According to Salem (2016), this model was developed after considering the needs and norms of the 180 nationalities living in Dubai. Salem further adds that this model is not biased to societal or demographic preferences. The possible benefits of developing a smart city are impartial to the origins of its residents. In other words, a hierarchy of beneficiaries of Smart Dubai is inexistent for citizens, residents, or visitors.

4.2 Smart City Program in the Kingdom of Saudi Arabia (KSA)

Unlike the UAE, the smart city program in the KSA requires a government platform to effectively implement firm policies and factors suited for the state. The smart city program is still in stages of development. Sporadic research work of consulting companies could be traced in the KSA, which indicates the smart development of certain sectors. Sari (2012) reported that the kingdom is keen on developing a smart city model that is suitable for its cities. In this context, the Saudi Arabia General Investment Authority announced the development of several smart cities in various parts of the kingdom.

According to CISCO (2009), the KSA invests billions of dollars to build four new "economic cities," which are referred to as smart cities. This action aims to diversify the economy of the kingdom by moving the gross domestic product away from the hydrocarbon sector (oil accounts for 94% of the export revenue of the KSA). The report further states that KSA is keen on developing a new smart project in the kingdom:

1. King Abdullah Economic City
2. Medina Knowledge Economic City
3. Prince Abdulaziz Bin Musaed Economic City

4.3 Smart City Vision of Qatar

A literature review about the smart cities in Qatar shows that Qatar is at the early stage of implementing smart city initiatives. Trivial development occurs in the cities of Qatar. However, the country has a futuristic plan to establish a new smart city named Lusail City that can house 450,000 people.

According to Kaldari (2016), Lusail City will set the new standards for the development of smart cities in Qatar. According to Kaldari, Lusail is a significant development plan that covers an area of 38 km^2 and is composed of four islands and 19 residential, mixed-use, entertainment, and commercial districts. This city will provide a high-technology environment with wired and wireless communication networks to offer advanced services. The new city will depict the actual features of a smart city because of its cutting-edge command and control center (LCCC). LCCC is the heart of this smart city where the management and monitoring of all smart services will be conducted. Qatar is ambitious about this development because it is the first project that is planned and executed by the government of Qatar. The key design elements and considerations indicate that Qatar focuses on establishing the future city that will enhance the lifestyle of people and empower businesses by providing efficient and sustainable services using an integrated ICT infrastructure.

4.4 Smart City Initiative of Kuwait

According to report published in Arab News (2016), Kuwait is close to adopting a smart city plan in its new housing projects to rationalize energy usage. Kuwait plans to establish a smart city agenda in major housing projects, such as Sabah Al-Ahmad City and Nawaf Al-Ahmad City. The government charter of smart cities in Kuwait is not yet fully developed. However, Kuwait initiated the sustainable development of its new cities and intends to develop future planning strategies by integrating three key factors, i.e., economic, social, and environmental (ESE) aspects.

Parallel to this development, Kuwait collaborates with the government of South Korea to plan and build a new smart city. The new four-billion smart city called South Saad Al-Abdullah City is located 40 km west of the center of Kuwait (Gulf Construction 2016).

Kuna (2016) reports that Kuwait initiates an environmental project called Al-Barayeh smart city. “Barayeh” in Arabic means an open area or social space where family members and children in particular use to gather for social interaction. The project is located in southern Kuwait. The structure of this project is based on the ESE plans of the government. The project will provide housing for 85,000 people and will maximize alternative energy resources and mass transportation system, which are the two key elements in the city.

5 Comparative Analysis of GCC Smart City Programs

Section 4 of the current study discussed the smart city program, which was introduced by GCC countries, particularly UAE, Saudi Arabia, Kuwait, and Qatar. A cross analysis of these programs and models revealed that the government of Dubai (UAE) developed an extensive smart city model and consulted various experts to devise a viable plan for the current conditions of Dubai. The various factors listed in the Dubai smart city model are already implemented. The government is eager to implement the entire smart city model by 2021. Health and the use of renewable energy are considered sub-factors.

Other GCCs showed sporadic development in terms of smart city programs. These countries implemented various ongoing projects related to the establishment of smart cities. However, these countries do not seem to adopt a comprehensive smart city model. The smart city models in these countries could be established by consulting and developing firms that are working on smart city projects. These models will fit particular design conditions and cannot be replicated to other cities. The smart city concept is introduced as a modern lifestyle for new developments but not as a system of city management.

6 Recommendations for Smart City Framework in GCC

GCC countries are attempting to establish expectations from unique smart projects for existing and total smart cities. These countries strive to implement a successful smart city model that can promote economic growth and human welfare. However, not all GCC countries fully developed a smart city model. The literature review and informal interviews with regional experts of the current study suggest the following framework, which could be utilized to develop a comprehensive smart city model for GCC, as shown in Fig. 13:

1. Wireless City: Mobile services are the future. The partners of smart cities are developing a range of new and innovative services for mobile platforms while testing and evaluating new forms of urban wireless networks. The wide use

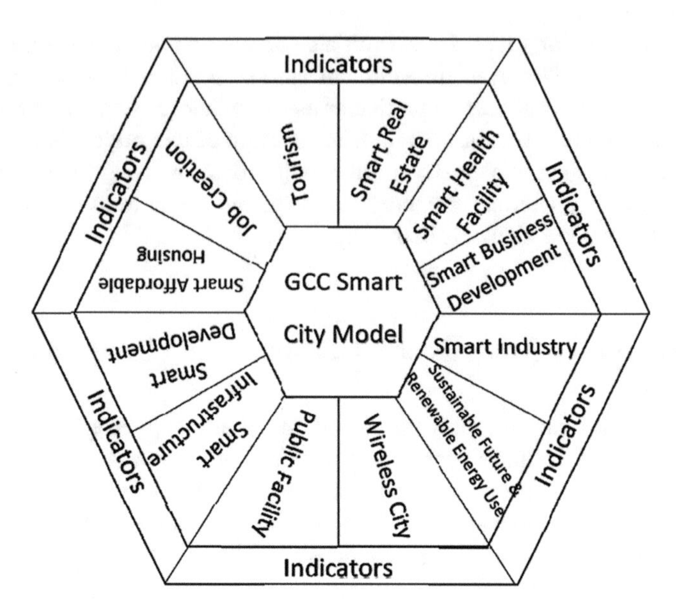

Fig. 13 Smart city model for GCC

of mobile phones and the emergence of municipal Wi-Fi networks allow local governments to deliver new services or adapt existing e-services to effectively reach citizens or workers on the move.

2. Public e-Facility: Nationwide adoption of e-government measures.
3. Smart Infrastructure: Developing a smart ICT-based infrastructure system for residents, which includes road networks, navigation, water and sanitation services, and electrical supply. This process can focus on increasing the efficiency of a city.
4. Smart Development: Smart cities enhance the developmental activities of a region. Various developmental activities, such as establishing schools, organizations, and shopping malls, can occur. These activities benefit all stakeholders, including citizens, businesses, governments, and the environment.
5. Smart Affordable Housing: Housing is major constituent of any smart city. Various factors, such as decent living standards, climate, peace, safety, and tax exemption, made GCC countries top settlement choices. The increased influx of people from around the globe requires additional housing. To address this concern, governments could set a prototype model of smart housing to provide ease and increase the efficiency of citizens.
6. Job Creation: A smart city can also serve as a cluster of an economy. This city provides various opportunities and advantages to its residents. Given the rapid expansion of GCC, a smart city can provide employment for many residents.
7. Tourism: GCC countries thrive on post-petroleum economies, but they are shifting toward a tourism-based economy. GCC countries aim to use their smart cities as tourism hubs to generate income.

8. Smart Real-Estate Retail: Certified real-estate retailers could be integrated through ICT platforms. Travelers and citizens can then seek help from this platform to identify solutions for living or business concerns.
9. Smart Health Facility: Health facilities are important for people of all ages, particularly the elderly. A network of health and care facilities could be integrated through an ICT platform, and users can utilize suitable and accessible health facilities according to their health concerns.
10. Sustainable Development and Renewable Energy Use: City managers can outline all indigenous or group development activities to ensure that smart features are integrated in all new development projects. These frameworks could be defined as partial-smart, typical-smart, and full-smart development. These frameworks could also suggest the best and tested mode of renewable energy for new development.
11. Smart Business Development: ICT can play a major role in establishing or managing existing businesses. City managers can schedule and streamline techniques and mechanisms to manage businesses from remote or offshore locations.
12. Industrial Growth/Smart Technology: Significant technologies and expertise are required to establish smart systems. Governments can regulate policies to facilitate research, production, and technology transfer of smart systems or mechanisms in a country or region. This initiative will result in industrial growth and will create jobs in a region.

7 Conclusions

The social setup of GCC countries is unique because of the presence of 180 nationalities in this region. This structure establishes the region as a harmonized multiracial hub. This unique feature helps this region maintain its economic growth. Despite the negative economic growth in several countries, the economic activities of GCC continue to increase, which result in new megaprojects. People from around the globe, particularly Southeast Asian countries, consider this region as a safe and secure destination for investment and world-class living. Cross-border urbanization is prevalent and offers many positive effects. Cross-border urbanization increases economic, business, and cultural activities and gathers people and ideas. However, an economic shift is also occurring in the region. Therefore, GCC strives to maintain its economic performance and switches from a petroleum-based economy to other successful income-generating modes, i.e., tourism, real-estate development, manufacturing, science and technology development, and exhibition.

The growing cross-border urbanization in GCC also causes problems, such as high population densities, urban mobility, and limited clean water, fresh food, and energy supply. To improve the quality of life and increase economic productivity, GCC cities should adopt systems that promote the efficient and sustainable use of resources. Ecologists and city experts believe that a city with well-planned ICT features offers various benefits because resource consumption for living is lowered, fuel usage for transportation is less, urban land is efficiently utilized, city governance and communication are effective, and income is saved.

Regional governments in GCC consider smart city concepts to minimize their fiscal deficits because these concepts can overcome stressful economic situations. The advantages of a smart city are multifaceted, which include environment, energy usage, transportation, communication, e-governance, and lifestyle. Following the efforts of Dubai, some GCC countries applied smart city concepts at various levels, such as by developing smart cities and implementing smart tools in existing. However, this region has yet to improve or establish fully developed smart city models.

The concept of smart cities provides a new wave of ICT technologies and convenience to citizens in GCC countries. Establishing such cities facilitates a clean environment, less driving, decreased number of cars on roads, decreased urban congestion, time and money conservation, increased time for family and health activities, easy governance, one-window operation, increased jobs, production of new industries, technology transfer, technology trade, and growth in economic activity.

References

Arab News: GCC plans to build smart cities. English daily. 21 February 2016. Visited 8.10.2016- http://www.technicalreviewmiddleeast.com/construction/buildings/gcc-plans-to-build-smart-cities (2016)

Caplescu, O.A.: Architecture in Science Fiction Movies, Romania, "Ion Mincu." University of Architecture and Urbanism (2012)

CISCO: Engagement Snapshot. Cisco Internet Business Solutions Group (IBSG) (2009)

Dassani, N., Nirwan, D., Hariharan, G.: Dubai -a new paradigm for smart cities. KPMG International Dubai 2015 (2015)

Department for Business, Innovation and Skills: Smart city market opportunity for UK. Research Paper No 136. 1 Victoria Street, London SW1H 0ET (www.gov.uk/bis) (2013)

European Commission: Smart Cities for Sustainable Development. Perspectives & Horizon 2020. University of Thessaly, Stavroula Maglavera (2014)

Gulf Construction: South Korea to build smart city in Kuwait. Visited 10-05-2016 http://www.gulfconstructiononline.com/stories/source/?url=/IND_306635.html (2016)

Hanson, M.: The Science Behind the Fiction – Building Sci-Fi Moviescapes, p. 72. Rotovision SA, East Sussex, 2005 (2005)

International Data Corporation: What is a Smart City. Infographic. IDC http://www.idc.com/home.jsp (2015)

Kaldari, E.M.A.: Future Smart City of Qatar. Lusail City Qatar. Visited 8.10.2016 http://www.lusail.com/ (2016)

Kuna: Smart city environmental project. English Daily. Kuwait Times. 08 February 2016. Visited 8.10.2016 http://news.kuwaittimes.net/website/al-barayeh-smart-city-project-wins-first-rank-in-gulf-region/ (2016)

MicroTAC: Motorola MicroTAC. Computer networking and telecommunication research http://www.cntr.salford.ac.uk/comms/etacs_mobiles.php (visited 21-9-2016) (1989)

Raitt, D., Warmbein, B., Edward, E.: Science fiction technology fact. ESA Publications BR-205 (2004)

Salem, F.: A Smart City for Public Value: Digital Transformation through Agile Governance – The Case of "Smart Dubai". Governance and Innovation Program, Mohammed Bin Rashid School of Government, World Government Summit, Dubai (2016)

Sari, A.: Smart Cities in Saudi Arabia – Current and Future. Japanese-Saudi Business Opportunities Forum The 12th Saudi-Japan Business Council February 1–2, 2012. Visited 10-05-2016 http://www.saudiarabia-jccme.jp/forumpdf/14-4.pdf. (2012)

Smart America: Smart Cities USA. http://smartamerica.org/teams/smart-cities-usa/ (2014)

Startrek: Communicators. Star Trek Movie. The classic American science-fiction/adventure television series. 1966. http://www.collectiondx.com/series/star_trek/star_trek_original_series_196669 (visited 21-9-2016) (1966)

UNECE: Smart urban solutions for transition and developing countries. Housing and Land Management Unit. United Nations Economic Commission for Europe (2015)

United Nations Department of Economic and Social Affairs: The World Population Prospects: 2015 Revision. 29 July 2015, New York. Visited: http://www.un.org/en/development/desa/publications/world-population-prospects-2015-revision.html (2015)

Zhao, H.: Accelerating the development of Smart Sustainable Cities Houlin Zhao, International Telecommunication Union News. No: 2. 2016 (2016)

Smart Cities: Distributed Intelligence or Central Planning?

Hermann Haken and Juval Portugali

Abstract We discuss various aspects of the application of AI/IT to cities, based on Portugali's concept of self-organizing cities, Haken's *Synergetics* as theory of self-organizing complex systems, and in particular on the Haken and Portugali study of the interplay between Shannon information and pragmatic/semantic information in cognition. In terms of allometry, it is shown that increasing automatization of a city may lead even to an increase of load on humans and in extreme cases to a system's instability.

1 Introduction

> Cities were always smart. In every era, advanced technologies and innovative thinking have developed in cities; from the written word 5,000 years ago; to the revolutionary Greek concepts of democracy and citizenry; to Renaissance art and architecture; to the factories of the industrial revolution; to today's post-industrial age of high technology. (Portugali 2016a)

This leads us to the question whether there is a specific feature that distinguishes the present days' concept of a "smart city" (Batty et al. 2012) from those of history. We think that a clue to answer this question is a look at our present most frequent use of the word "smart." In fact, we speak of smartphones but also of smart houses, smart households, smart cars, etc. At a still larger scale, smart factories are conceived. These smart objects are related to the notion of *the Fourth Industrial Revolution* (Schwab 2016) suggesting that today once again society is at the threshold of an industrial revelation: the first happened in the eighteenth century by the introduction of mechanical machines, the second in the second half

---

H. Haken (✉)
Institute for Theoretical Physics, Center of Synergetics, Stuttgart University, Pfaffenwaldring 57/4, D-70550, Stuttgart, Germany
e-mail: haken@itp1.uni-stuttgart.de

J. Portugali
ESLab (Environmental Simulation Lab), Department of Geography and the Human Environment, Tel Aviv University, Tel Aviv, 69978, Israel
e-mail: juval@post.tau.ac.il

S.T. Rassia, P.M. Pardalos (eds.), *Smart City Networks*, Springer Optimization and Its Applications 125, DOI 10.1007/978-3-319-61313-0_5

of the nineteenth century via electrification. The third revolution is characterized by computers and microelectronics, while presently we witness the rise of smart factories by what is called "digitization" of a network comprising construction, development, production, sales, and services (cf. Bauernhansl 2015). This implies an integration between real and virtual worlds, enabling the simulation of systems, processes, and even complete factory plants in real time. Here, the reconciliation between long-term planning and short-term reactions to customers' wishes and market fluctuations presents a real challenge. Planners presently consider the man-robot cooperation from the point of view of a combination of the cognitive superiority and flexibility of humans and the power, endurance, and reliability of robots. Quite evidently, the concept and development of smart cities will present a far greater challenge that has to take care of the central task of a city – the welfare of its citizens.

The currently developing discourse on smart cities is intimately associated with complexity theories of cities (CTC) – a domain of research that in the last decades applied the various complexity theories developed since the early 1960s to the study of cities. Smart cities with their massive use of artificial intelligence (AI)/information technology (IT) are considered as one central way to cope with the growing complexity of cities, while the set of urban simulation models developed within the context of CTC is considered one among the new information communication technology (ICT) that enable the smartification of cities (Batty et al. 2012).

Recent studies in the domain of CTC indicate that while cities share many properties with natural complex systems, they differ from the latter in that they are *hybrid complex systems* composed of artifact that are by definition simple systems and of human urban agents that are natural complex systems (Portugali 2011, 2016b). In today's cities there is a clear distinction between the artificial components of cities as hybrid complex systems (e.g., houses, streets, etc.) and their natural components – the human urban agents. Artifacts cannot interact (exchange things/information), agents can. Artifacts interaction is thus *mediated* by agents. Also, a lot of agents' interaction is mediated by artifacts (as treated by the SIRN approach of Haken and Portugali 1996). With AI/IT, artifacts might be able to interact directly without mediation by human agents and thus become urban agents. For example, a self-driven car can interact with a self-organized traffic light in a junction. Artifacts thus become urban agents giving rise to a new form of urban dynamics. Part of it may be smart traffic regulations (guiding system); smart supply systems of energy, food, and goods; or smart waste disposal. But nevertheless, who or what is smart?

The technical installations (houses, communication systems, supply systems), their human planners and users, or their interplay? At any rate, we are facing a delegation of human decision-making and responsibilities to automata. So all in all, we believe that "smart city" implies a qualitative change based on artificial intelligence (AI) embodied by information technology (IT) on all scales. This means that we must not ignore the fact that cities are embedded in nations in an increasingly connected world which is becoming also "smart" using AI (just think of the financial market with its computer-controlled transactions). At all levels, there will be a demand for sensors, actuators, and computing power. The wide use of AI/IT devices will lead to an innovation wave quite welcome to economists because innovations are seen as a motor of economic growth and public welfare. In the following we want to elucidate some aspects of "smartification."

Basically, we may distinguish between two approaches: the top-down approach, where data are collected locally and sent to a central computer, which makes the decisions, or a bottom-up approach where the decisions are made at the individual level based on collected data. The latter approach is outlined in a recent study by Feder-Levy et al. (2016), based on Portugali's concept of self-organizing city (Portugali 2000, 2011, 2012). The theory of synergetics introduced below offers an integrative approach: local bottom-up decisions and actions give rise to a collective structure that then in a top-down manner determines ("enslaves" in the parlance of synergetics) local actions and decisions and so on in circular causality.

In the beginning, solid structures, such as houses and streets, will not be changed. Only at a later stage, changes may become necessary due to a newly developing dynamics of traffic, but also of personal habits. For example, IT may lead to a delocalization of teaching and, perhaps, education, by replacing schools and universities by telecourses, e.g., massive open online courses (MOOC). Because this implies a loss of personal contact between teachers and students (and among students), it is, however, unlikely that schools and universities will disappear completely. The same holds true for sports – or cultural centers. Can, e.g., in a baseball or soccer game virtual reality replace the feeling of being member of some community?

1.1 On the Interplay Between Humans and Smart Devices

Smart devices on nearly all scales will replace human senses and (re)actions. A few examples may illustrate this. In a smart home, sensors may measure the actual amount of daylight/sunshine, rainfall, in- and outdoor humidity and temperature, energy consumption, etc. In a city, sensors may measure local traffic flows, local and overall energy consumption, etc. So far in a home, individuals have reacted to the data, e.g., temperature according to their specific habits. Now, "AI devices" learn all these human reactions, e.g., to close the venetian blinds at a certain level of sunshine. But there may be conflicts between family members. While A wants them to be closed, B wants them to remain open. This learning problem can perhaps be solved by some kind of majority decision based on relative frequencies of action. At any rate the AI device will then act instead of a person. This is surely convenient for the individual but leads also to a reinforcement of his/her habits, i.e., his/her habituation.

From the point of view of the complexity theory of synergetics (cf. Sect. 4 below), the above situation implies that the AI program has become the order parameter (OP) that now, in the literal sense of the word, enslaves the individuals, the family. Having the properties of OPs in mind, it will be difficult to change that

OP. Such habituation effects have been repeatedly discussed in connection with advertisements based on consumers' behavior. At the political level, even some kind of nudging has been discussed.

A down-to-earth technical problem should be mentioned; when a wireless local area network (WLAN) is used in neighboring flats/homes, interference effects may spoil the operations. Both *conflicts* of interests among citizens and habituation may occur at the city level: chosen car routes (think of autonomous cars interacting with traffic guidance systems), total energy consumption, use of communication channels, supply routes for commodities, etc.

While in this way, the former more or less self-organized collective behavior of the citizens will be learned by the AI system and "regulated" correspondingly, we may think of quite another scenario: a central city computer tries to solve a multi-traveling salesman problem. Though we don't know how such an approach would look like or what its results/efficiency will be, such a scenario is by no means unlikely. This may perhaps lead to large computer centers outside cities where building sites are cheap. Finally, centralized installations will increase the vulnerability of cities against crime, terror acts, and global breakdown. Another basic problem is, e.g., energy consumption control. A typical "recipe" runs like this: increase/decrease of energy price over the day/night depending on consumption, thus influencing the consumers' action. As is well known, this scenario may lead to instabilities in the network. A stricter central control may eliminate the instabilities, but will curb individual freedom: clearly, smartification has sociological implications.

1.2 Theoretical Tools

To study implications of "smartification" from the point of view of theory, we have a number of approaches at hand, i.e.:

1. Cognitive science
2. Artificial intelligence
3. Information theory
4. Synergetics
5. Network theory
6. Multi-agent theory
7. Evolutionary game theory
8. Allometry/scaling laws
9. Biology – evolution, population dynamics

Besides these approaches with their roots in mathematics and the "natural" sciences, other disciplines will play a role in smart cities such as:

(a) Jurisprudence (e.g., responsibility transfer, liability)
(b) Sociology (e.g., job market requiring highly qualified personnel for IT-maintenance, development of AI/IT, but also disappearance of other jobs, e.g., taxi drivers, bank clerks, etc.)

(c) Psychology (e.g., psychological stress in an automated world)
(d) Political science (e.g., decision-making on cities' smartification in a democratic society)
(e) Economics (e.g., investments in smartification)
(f) Ecology (e.g., management of resources by smartification)

Having said this, we focus our attention on some of the topics 1–9. First we briefly discuss them. Since, at least according to our understanding, "smart city" implies a massive use of AI/IT, a discussion on intelligence in general and on AI in particular may be in order. Here, we have to draw on insights gained by cognitive science. We will elaborate this in Sect. 2. In this context, we will briefly discuss information theory and its more recently established connections with cognition such as *information adaptation* (Sect. 3).

Cities with their inhabitants and mobile and immobile installations (artifacts) are truly complex systems that form specific spatial and functional structures where the interplay or competition between central planning and local personal initiative becomes quite decisive (Portugali 2011, 2012). A general theory of structure formation in complex systems is provided by *Synergetics* that we will outline in Sect. 4. This will shed light on the role of indirect steering in contrast to conventional planning. Network theory (cf., e.g., Barabási 2016) gives insight into the links, e.g., between inhabitants or between neighborhoods or cities, and allows, e.g., the derivation of scaling laws. On the other hand, multi-agent theories (cf., e.g., Bretagnolle et al. 2006; Roscia et al. 2013) deal with the actions of citizens and automata. Quite clearly, network theory will play an important role in IT dealing with communication and transport. In the context of our paper, we are rather concerned with AI aspects. Evolutionary game theory (EGT) deals with the benefits of cooperation between partners (persons, companies, institutions, etc.). The concept of Nash equilibria (well known in EGT) will play a role in our discussion. EGT originated from game theory developed by von Neumann and Morgenstern (1944). The biological concept of evolution, in particular based on *mutations*, has become an important ingredient of EGT (e.g., Nowak 2006). Mutations are treated as chance events, also denoted as *fluctuations*. As is shown quite generally in *Synergetics*, fluctuations play a fundamental role in the self-organized formation of structures. Fluctuations can be also considered as means to test the stability or resilience of a system against perturbations and to trigger novel developments. A further line of research inspired by biology is *allometry* which is being applied to theories on city growth. In biology it was found that e.g. the age, size, blood flow and other characteristic features of animals scale with body weight w by means of a power law, $\propto w^{l/4}$, $l = 1, 2, 3$ (West and Brown 2005). Similar laws have been considered in the case of cities (Bettencourt et al. 2007). We will discuss such laws with respect to "smart city" at several instances below (Sects. 5 and 6).

2 Intelligence

Having in mind that "smartification" implies the application of AI to cities, a few remarks on intelligence may be in order, at least what its role in cities concerns. Most probably, it is sufficient to deal with *intelligent behavior*. This in turn means appropriate *actions* or, depending on the specific situation, *reactions*. A prerequisite for these processes is *pattern recognition*, where "pattern" may be interpreted in a wide sense, e.g., as images of faces, or objects, or whole scenes, or movement patterns of persons, groups of them, of traffic, or just a set of measured data, e.g., temperature distribution in a home or in a city. The central problem consists in the selection/recognition of features that are/will be relevant for (re)action and to draw the relevant conclusions. To have the required actions *quickly* at hand, a whole repertoire of (re)actions must be learned in advance. A number of them may be based on planning, and foresight is required. An important problem is the challenge of new situations to decision-making. In many cases, people base the latter on more or less justified analogies with previous experience by extrapolating ("heuristics"). But in a number of cases, entirely new solutions (actions) are required. Among such chance events are technical failures, human mistakes, natural catastrophes, new state laws, etc.

So far we have spoken of intelligent behavior of an individual. But there is also collective intelligence. The famous economist Friedrich August von Hayek spoke of "intelligence of the market" which in the present context can be extended into the "intelligence of the city." Some collective intelligence is even attributed to swarms of birds or schools of fish. Quite a number of scientists think that collective intelligence cannot be substituted by individual intelligence. At least one of the reasons is the "information bottleneck": a central agency, or even an individual, cannot deal with the huge amount of incoming information. Thus only delocalized decision-making (by the market) is possible. Clearly, there is a dichotomy between local initiative and global planning. Coming back to "smartification," taken seriously, it means the replacement of human intelligence – as sketched above – by artificial, that is, machine intelligence.

2.1 Artificial Intelligence (Machine Intelligence)

Here we don't discuss philosophical issues such as the "ghost in the machine" nor ethical, such as responsibility or liability of machines (cf., e.g., Bonnefon et al. 2016).

The first step consists in the collection of data by sensors for optical, acoustic, chemical, tactile, acceleration, etc. signals. Present days' data may be enormous, hence the notion of "big data." If not appropriately preselected ("filtered"), they mirror a complex world. The next, in our view, decisive step is supervised learning (cf., e.g., Le Cun et al. 2015). For example, in image recognition, a huge number of faces of the same person but in different positions, under different illuminations, are presented to the computer that has to "learn" that all these images are *associated* with the name of that person. This is accomplished by an algorithm with a huge number of adjustable parameters. This procedure is usually visualized as a feed-forward net with a considerable number of layers (say 20–100). Most of the adjustment procedures are based on the method of "steepest descent." To get an idea what this means and implies, think of a landscape with mountains and valleys in between. The bottom of each valley represents a specific prototype pattern to be recognized due to given data which are represented by the position of a stone in this landscape. Since in general the data are incomplete and/or partially erroneous, the position of the stone does not precisely coincide with that of the bottom of the relevant valley. But the stone can correct this error by sliding downhill.

This little side remark may shed light on a basic problem: how certain is it that the "envisaged" valley is the correct one? This also implies that under somewhat changed conditions, the previous valley is no more the appropriate one if it has happened so before. This means there is no guarantee for the adaptability or generalization capability of this algorithm. Besides such a rather superficial consideration (and related ones), there is no theory of such networks that would allow us to understand the learning process. In a way the whole learning procedure is reminiscent of the training of animals. We don't know what happens in their brains. And perhaps, once the tiger bites, quite unexpectedly. An additional remark may be in order: at least at present, the training time of such a network is days, which may be shortened by the parallel use of thousands of computers. Nevertheless, the speech recognition capability of systems such as Siri (Apple) is impressive. So far we have talked about recognition (which in case of AI may include that of scenes and of actions of persons). What about action of automata? The same way as recognition can be "taught" to a computer, it can also be taught to steer movements of actuators in particular by *imitation*. But then, quite often, the machine is confronted with a conflict situation: which movement to steer (perform)? The inability of the machine to make an autonomous decision leads to deadlock. In human life, in such a situation, a deeper insight, or deeper experience, is required and helps to overcome the deadlock. On the other hand, if such a conflict situation hasn't been "shown" to the machine, deadlock results.

3 Information Dynamics

Clearly, the concept of information plays a fundamental role in IT. However, when applying this notion to urbanism, we have to exercise some care. The reason is that the concept of "information" used in IT is quite different from the use of this word in regular language. In the latter case, "information" may denote a message, such as "Bob arrives today," or an order such as "close the door!". In these cases we deal with semantic or pragmatic information, i.e., information carrying *meaning*. In IT, conventionally, "information" is used as defined by Shannon in his fundamental paper (Shannon 1948):

$$S = -\sum_j p_j \log_2 p_j,$$

where p_j is the probability distribution of events distinguished by an index j. According to this definition, Shannon's information (SHI) doesn't carry any meaning which allows its wide range of applications. On the other hand, the fundamental question arises, whether there is, nonetheless, a relation between Shannon information, pragmatic information, and semantic information. Aside from earlier work, e.g., Floridi (2011, 2015), this question has been recently dealt with by Haken and Portugali (2015, 2016) who call the interplay between Shannonian and semantic/pragmatic kinds of information, *information adaptation (IA)*. Its basic idea is as follows (for a detailed mathematical approach, cf. Haken and Portugali 2016): according to Shannon's formula, SHI depends on the selection of events j. The selection depends on the human sensorium, e.g., acoustical, optical, tactile, etc. as well as on hypotheses, e.g., on expected correlations between stimuli and even on possible interpretations. As we termed it, "semantic information enters Shannon information in disguise." The thus preprocessed SHI is then transferred – by the human brain or, in the context of "smartification," by an automaton – into a state of the brain/automaton associated with meaning, which can refer to semantic information (SI), that is, meaning per se (e.g., this is "sky") while in particular to soliciting a specific action ("pragmatic information" (PI)). In a human brain, this interplay between SHI and PI may happen in several steps. Depending on the amount of SHI (itself depending on preselection by means of SI/PI), a *unique* response ("information deflation") or a multitude of responses ("information inflation") may result.

After our introductory remarks, we now turn to a study of the information dynamics in the city. Our approach must be considered as a first step toward a more comprehensive and detailed theory. We base our approach on the distinction between SHI and PI. (Here we omit a discussion of the difference between semantic information (SI) and PI which would distract our attention from our basic goal.)

Let us ignore, in a first step, the role of automata and focus our attention on the role of humans ("agents") who play a double role: they receive information and produce information. But what kinds of information are meant here? To cast our approach into a mathematical form, we assume that the received information is of Shannon type and the "produced" information of pragmatic type. The latter leads to observable actions of the agents. As we show in some details in our IA study (Haken and Portugali 2015), because of their recognition capabilities, the agents transform the data emitted by the environment firstly to SHI and then into PI. For simplicity, in what follows, we'll deal with the conversion of SHI into PI. Let us consider the role of the city that is composed of artifacts and agents. Both emit signals (e.g., the artifacts in Gibsonian language as affordances). All these signals are treated as "raw material" in form of SHI that has then to be "deciphered" by the agents, i.e., converted in PI.

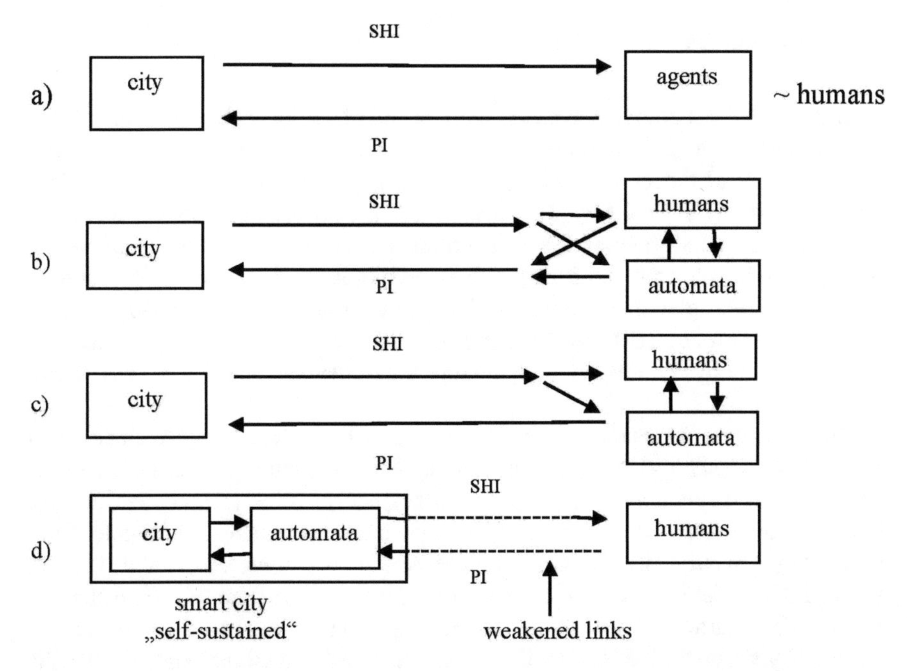

Fig. 1 Stages of smartification (cf. also text). In the case of automata dominance $a^2 r_a r'_a / \gamma_a \gg n^2 r_h r'_h / \gamma_h$ (cf. Sect. 6)

Thus SHI is the number of bits produced by a city, e.g., per day, and embodied by letters, on monitors – a gigantic, but nevertheless, measurable number. These signals are received, filtered, and interpreted by the human sensorium and eventually converted into PI. This leads us to our model depicted in Fig. 1a. As can be seen, Fig. 1 illustrates graphically four stages in the smartification of a city, starting from Fig.1a, representing the current state of cities, all the way to Fig. 1d that represents a "fully smart" city as discussed in more details, in Sects. 5 and 6 below. The four stages in the smartification of a city are:

(a) The city and its human inhabitants exchange information. While the city produces SHI, its human inhabitants recognize these signals and convert it via PI into actions.
(b) Automata participate in the exchange process.
(c) The direct action of humans on the city has practically disappeared and is mediated by automata.
(d) Automata are perceived by humans as part of the city.

4 Synergetics

A city is a hybrid complex system "composed" of its inhabitants ("agents") and mobile and immobile installations that all are strongly interwoven in different ways. A general theory of the behavior of multicomponent systems is provided by *Synergetics* (Haken 2004). This field of research deals with the self-organized formation of spatial, temporal, and functional structures and has found numerous applications in fields ranging from physics over biology and psychology to problems in city dynamics, planning, and design (Portugali 2000, 2011: Portugali et al. 2012: Portugali and Stolk 2016). Here is a brief outline of its basic concepts, because the insights gained by synergetics are quite relevant to the question: "local initiative or central planning"?

The basic concepts and research strategy of synergetics are, roughly speaking, as follows. A *system* (city) is composed of mutually interacting *elements*. The elements are subject to specific *control parameters* (taxes, energy prices, rental rates, costs of land) that are either fixed externally (e.g., by state government) or internally (e.g., mayor/city government). These control parameters are considered fixed, i.e., not reacting to the elements during the considered time interval. It is assumed that under fixed control parameter values, the system has found a *structural stable state* (a special city structure that is spatial, e.g., street network, and functional, e.g., traffic, energy supply, etc.). *Structural stability* means that the system is able to return to its former state after a *disturbance* (e.g., fluctuation of size of population such as immigration or emigration, temporal breakdown of electricity supply, or, locally, car crashes). Synergetics focusses its attention on situations where *stability* is *lost*. This may happen when some control parameters are changed (e.g., taxes, rental rates, and also human desires (see below)). Under these new circumstances, the system tends to form a new structure (e.g., segregation of specific populations). Based on a detailed mathematical approach, Synergetics has provided us with quite remarkable insights.

1. Close to an instability, in general, *several* new structures become possible (e.g., the newly developing living quarters may be differently located).
2. Which structure is determined depends on a "chance" event (e.g., the initiative of an individual or a small group of people).
3. The behavior of the many individual components (e.g., citizens) is governed by very few collective variables called *order parameters* (e.g., commonly shared ideas, concepts, beliefs, etc.). They may be used as descriptors of the global situation and obey their own dynamics.
4. Most importantly, *order parameters* (OPs) determine the behavior of the components. This is called the *slaving principle*. Note that this is purely a terminus technicus without any ethical implication.
5. The concept of OPs allows an enormous *complexity reduction* because instead of dealing with the numerous components, we have to treat only the behavior of few OPs.

6. A somewhat "philosophical" question concerns the notion of *circular causality*: while the OPs are brought about by the collective behavior of the components, the behavior of the latter is "enslaved" by the OPs.

These results allow us to draw some general conclusions provided that we deal with *self-organization*. This means that the structure of a system ("structure" in a very general sense) is not imposed on it from the outside (like a statue is formed by a sculptor) but evolves in response to rather unspecific conditions represented by control parameters. In a number of cases, we may speak of adaptation which implies that, e.g., a biological system enhances or secures its survival chance, but "self-organization" may also lead to a disaster, e.g., extinction of a population. The self-organized formation of a structure requires the interplay between

(α) Instability
(β) Chance event

as an important tool of influencing complex systems. While in commonly known *direct steering* each agent is precisely instructed how to behave in a specific manner, in indirect steering, the agents are informed only about the change of one or few general conditions (e.g., an increase of the price of building sites, rental rates, interest rates, or even new traffic regulations (e.g. speed limits). It is then left to the ensemble of agents how to respond to these new conditions. Typically there are two kinds of responses:

(a) The agents react smoothly so that the global structure is only slightly distorted.
(b) Because of the agents' responses, several qualitatively new global structures may result which under the given conditions are equivalent. Thus in the jargon of *Synergetics*, point (α) leads to the concept of *indirect steering*. The system's components are not instructed how to behave, but only global control parameters must be fixed. In view of the multitude of outcomes, a crucial question concerns the selection among these outcomes. This can be achieved in two ways.

    1. By an intervention external to the system. In a number of cases, even a slight push may suffice to let the system make the wanted choice. Strictly speaking, in this case, the system is no more self-organizing. For instance, in a regulated economy, the prices of building sites can be locally adjusted.
    2. By an internal chance event (β), such chance events are well known in physics as thermal or quantum fluctuations and serve as trigger for phase transitions. But they play a fundamental role also in social systems.

A nice example for the interplay between α) and β) is provided by the "balconies of Tel Aviv" as reported by Juval Portugali (2011). In terms of synergetics, this process can be interpreted (or explained) as follows: the relevant control parameter is the desire of many people to enlarge their flats or, in other words, to change the present state. Thus the system is ready for a change – its state has become *unstable*. The required chance event is the action of a single inhabitant who closes his/her balcony. And the relevant OP is the number of now closed balconies.

Quite evidently, taking the mechanisms of self-organization seriously, the appropriate choice of control parameters is a crucial task for any planning agency: how to induce the relevant agents to enable the formation of a new structure? EGT, stressing the importance of benefits via cooperation of agents, is an appropriate tool. Similarly to any organism, a city too must be adaptive so to meet changing needs and changing demands for labor, supply, etc. As has been noted, e.g., in ecosystems (Holling 1986), but also quite generally in synergetics, there a dilemma arises: the more stable a system is, the less adaptive it is and vice versa. As we will show elsewhere, stability and adaptability are related by the formula

$$\text{stability times adaptability} = 1.$$

This requires that organisms as well as cities to be viable must be kept close to instability or brought to it at appropriate instances.

5 Information Dynamics and Allometry in Smart Cities

In this section we apply the insights we have gained in our previous work (Haken and Portugali 2015, 2016) on the interplay between Shannon information (SHI), pragmatic information (PI), and semantic information (SI) to the dynamics of information in smart cities. In particular, we focus our attention on the role played by automata. Our basic approach is depicted in Fig. 1. We distinguish between three producers and processors of information: the city as a whole comprising artifacts (buildings, streets, etc.), human agents, and automata (artificial agents). We treat the city as a sender of information (where the role of artifacts is interpreted in Gibson's sense). The emitted information is treated as SHI and measured in bits. Its recognition is left to the receivers, i.e., to humans and automata. Thus according to our previous analysis, SHI is converted into PI (and SI). PI becomes visible as specific actions performed by humans and automata. Actually, pattern recognition capabilities are ascribed to the latter, as well as the capabilities of appropriate reactions. We include SI (semantics) in PI, provided ideas, concepts, plans, etc. ("mental states") are externalized, e.g., by written texts; computer programs, including those for graphics; etc. (see also SIRN – Haken and Portugali, 1996). We measure PI in bits. Since in our approach we are dealing with gross features to illuminate the impact of automata on information, we consider the total amount of SHI and PI per 24 h in a city. We assume an unlimited channel capacity. Clearly, more IT-oriented approaches will have to consider the effect of channel capacity also. We assume that the conversion of SHI$\rightarrow$PI occurs within 24 h ("temporal course graining"), and we ignore the local fine structure ("spatial course graining"). We treat the city as an open system with imports and exports of raw material, commodities, ideas, etc. The variables of our basic equations are

SHI produced per day by the city, denoted by s
PI produced per day by humans, denoted by p_h
PI produced per day by automata, denoted by p_a

n is the number of citizens
a is the number of automata.

In the spirit of *Synergetics*, s, p_h, and p_a are the order parameters, while n and a act as control parameters jointly with rate constants defined in the following.

We come to the formulation of rate equations

p-rates

(1) Generation rate of p_h (by human actions)

$\left.\frac{dp_h}{dt}\right|_1 = n \cdot s \cdot r_h$: humans transfer incoming Shannon information into PI_h (actions) at rate r_h

(2) Generation rate of p_a by automata

$\left.\frac{dp_a}{dt}\right|_1 = a\ s\ r_a$: automata transfer incoming Shannon information into PI_a (actions) at rate r_a

(3) Loss rate of p_h

$$\left.\frac{dp_h}{dt}\right|_2 = -\gamma_h p_h$$

(4) Loss rate of p_a

$$\left.\frac{dp_a}{dt}\right|_2 = -\gamma_a p_a$$

loss because of forgetting, executed actions by humans or automata, storage in memory.

s-rates SHI is produced by the city

(5) Generation rate of s

$\frac{ds}{dt} =$ generated data D

(6) By humans $D_h = n \cdot g_h + n r'_h p_h$, g_h: spontaneous generation rate per human,

$r'_h p_h$: stimulated actions per human

(7) By measurement devices D_a (automata) e.g. energy consumption, traffic flow

$D_a = a\ g_a + a r'_a p_a$, g_a: generation rate per automaton,
$r'_a p_a$: stimulated action per automaton

(8) Loss rate of s:

(9) By conversion $s \to p_h$: $\tilde{r}_h\ s\ n$ (via humans)

(10) By conversion $s \to p_a$: $\tilde{r}_a\ s\ a$ (via automata)

(11) By errors, $\gamma_e s$, where γ_e error rate

(12) By spam, chunk included in (11)

Total rate s

(13) $\frac{ds}{dt} = n \cdot g_h + \mathrm{a}g_a + n\ r'_h p_h + a r'_a p_a - \tilde{r}_h s\ n - \tilde{r}_a s\ a - \gamma_e s$
We consider the steady state

(14) $\frac{dp_h}{dt} = \frac{dp_a}{dt} = \frac{ds}{dt} = 0$
The sum of the generation rate (1) and loss rate (3) jointly with (14) yields

(15) $n\,s\,r_h - \gamma_h p_h = 0$

similarly, (2), (4), and (14) yield

(16) $a\,s\,r_a - \gamma_a p_a = 0$

the sum over (6), (7), (9), (10), and (11), jointly with (14) yields

(17) $n\ g_h + a\ g_a + n\ r'_h p_h + \mathrm{a}r'_a p_a - \tilde{r}_h s\ n - \tilde{r}_a s\ a - \gamma_e s = 0$

these three equations for s and p_h, p_a depend, in particular, on the parameters n (humans) and a (automata)

Solution to (17)

(18) $s = (\tilde{r}_h n + \tilde{r}_a a + \gamma_e)^{-1}\left(n\ g_h + a\ g_a + n\ r'_h p_h + \mathrm{a}r'_a p_a\right)$

Solution to (15) and (16)

(19) $p_h = \gamma_h^{-1} n\ s\ r_h$

(20) $p_a = \gamma_a^{-1} a\ s\ r_a$ from which we obtain

(21) $p_h / p_a = \gamma_h^{-1} n\ r_h / \left(\gamma_a^{-1} a\ r_a\right)$ or, equivalently

(22) $p_a = A p_h, A = \gamma_a^{-1} a\ r_a / \left(\gamma_h^{-1} n\ r_h\right)$ and with (18)

(23) $p_h = \gamma_h^{-1} n\ r_h (\tilde{r}_h n + \tilde{r}_a a + \gamma_e)^{-1}\left(n\ g_h + a\ g_a + n\ r'_h p_h + a\ r'_a p_a\right)$

(24) $p_a = \gamma_a^{-1} a\ r_a (\tilde{r}_h n + \tilde{r}_a a + \gamma_e)^{-1}\left(n\ g_h + a\ g_a + n r'_h p_h + a r'_a p_a\right)$

Inserting (24) in (23) yields

(25) $p_h = B(C + Dp_h)$ and thus

(26)

$$p_h = (1 - BD)^{-1} BC$$

B, C, and D are defined by

(27)

$$B = \gamma_h^{-1} n\ r_h (\tilde{r}_h n + \tilde{r}_a a + \gamma_e)^{-1}$$

(28)

$$C = n\, g_h + a\, g_a$$

(29)

$$D = n\, r'_h + a\, r'_a A,$$

or, with (22)

(30)

$$D = n\, r'_h + \frac{a^2}{n} \frac{\gamma_h}{\gamma_a} \frac{r_a r'_a}{r_h}$$

The solution (26) requires

(31) $(1 - \mathrm{BD}) > 0$
because $P_h > 0$ and $\mathrm{BC} > 0$.

$(1 - BD) = 0$ means that no steady-state solution exists in contrast to the assumption (14). In this case the complete time-dependent equations must be considered which lead to instability.

This instability is caused by a feedback loop SHI →PI→SHI inherent in our (time-dependent) equations. In the spirit of allometry, we are interested in the functional dependence of p_h (humans) and p_a (automata), i.e., human and automata activities, on parameters, in particular number n of citizens and number a of automata, and rate constants. In the conversion rates $r_h, r_a, \tilde{r}_h, \tilde{r}_a$ for each individual human or automaton, a number of preselection constraints enter, such as special kinds of transmission channels, their network structure, wired, wireless, etc. For our analysis it will be sufficient to discuss the relative size, at least of some pairs of rate constants. The rate r_h determines the information transfer from SHI (city) to PI (humans), while $\tilde{r}_h$ the information transfer from PI (humans) to SHI (city). Their relative size can be determined as follows. We consider these processes ignoring all other processes. Then according to (1) and (13), we obtain for their sum

(32) $\frac{d}{dt}(s + p) = (r_h - \tilde{r}_h)\, s$

$r_h - \tilde{r}_h = 0$ means that $s + p = const.$

Here we can draw on our previous work where we have introduced the notions of deflation and inflation meaning here

(33) $\frac{d}{dt}(s + p) < 0$ (deflation), $\frac{d}{dt}(s + p) > 0$ inflation.

(34) In practice, we expect deflation, which means $r_h < \tilde{r}_h$ (at least on average).
The same conclusion holds true for automata,

(35) $r_a < \tilde{r}_a$.

The size of $r_h(r_a)$ is a measure of the ability of humans (automata) to convert SHI into PI, i.e., to recognize signals and to convert them into action. At present, surely $r_h \gg r_a$, but the concept of a “truly” smart city might imply $r_a > r_h$. In view of the big progress made in computer linguistics, the experimental determination of the transfer rates r per human (or automaton) where r may depend on the finite range of included topics is in our opinion possible. The *r and s* we use in our Eqs. are average values over many humans (automata) relevant for city life (including “import,” “export”). Little can be said about the decay rates γ, but surely γ_e will increase in the course of time because of spams, etc. When dealing with smart cities, we must be careful when comparing the rates $r_h, \tilde{r}_h$ with $r_a, \tilde{r}_a$, because our use of the word automaton may comprise a wide range of interpretations – from smart household devices over robots till large computer centers. So when we consider special cases, we will have to take these distinctions into account. In particular, when computer centers play an important role, we have to compare the combinations $r_h n, \tilde{r}_h n$ with $r_a a, \tilde{r}_a a$. Finally we have to discuss the inclusion of export and import of material and immaterial goods. The effect of import (or “input”) can be taken care of by extending the interpretation of D_h (6), D_a (7). In (6), we may replace the spontaneous generation rate g_h by

(36) $g_h = g_{hc} + g_{hi}$

where g_{hc} is just the former spontaneous rate, whereas g_{hi} is the generation rate of SHI induced by imports. Similarly, we may proceed with (7), by making the replacement

(37) $g_a = g_{ac} + g_{ai}$

Note that, at least at the present state of AI, we may put

(38) $g_{ac} \approx 0,$

because we assume that AI devices can hardly produce truly novel ideas. The rate of export, E, can be expressed by $E = r_{he}p_h + r_{ae}p_a$ with the rate constants r_{he} , $r_{a,e}$. In our approach, the SHI, PI dynamics is not directly influenced by E, but indirectly via the loss rates γ_h, γ_a. While clearly our model refers to a whole city and all its activities, a number of similar models can be established to deal with specific issues. Just to mention a typical example: humans observe (“measure”) the walking speed of other humans (SHI) and respond by adjusting their own speed (PI) taking into account other relevant features of the city, in particular its size (e.g., Haken and Portugali 2016).

6 Special Cases Giving More Insight

In this section we discuss the dependence of p_h on the various parameters, particularly on n and a but also on the rate constants r. The special cases we treat are characterized either by the dominance of human activities over those of automata (case h) or vice versa (case a). Our starting point is Eq. 26 with its quantities B, C, and D (27–30). The just mentioned cases (h, a) can best be distinguished by means of D (30). Multiplying (30) by $n\, r_h \gamma_a$, we obtain on the r.h.s. a sum of

(39) (h) $n^2 r_h r_h' / \gamma_h$ that refers to human "attributes"
(40) (a) $a^2 r_a r_a' / \gamma_a$ that refers to automata "attributes"
According to (h) $\gg$ (a) or (h) $\ll$ (a), we derive simplified expression for p_h.
First we assume
(h) $\gg$ (a)
D (30) reduces to $D = n\ r_h'$, while B (27) reduces (under the assumptions $\tilde{r}_h n \gg \tilde{r}_a a + \gamma_e$ and $\tilde{r}_h \approx r_h$) to

(41) $B = \gamma_h^{-1}$

so that

(42) $BD \approx n \frac{r_h'}{\gamma_h}$
Putting (26), (41), and (28) together, we obtain for the "load" per person

(43) $p_h/n \approx \left(1 - n \frac{r_h'}{\gamma_h}\right)^{-1} \gamma_h{}^{-1} \left(g_h + \frac{a}{n} g_a\right)$

where the impact of the term $\frac{a}{n} g_a$ is, in the frame of our approach, negligible.

Most remarkable is the factor $\left(1 - n \frac{r_h'}{\gamma_h}\right)^{-1}$ which causes an enhancement of the spontaneous "production rate" g_h. This factor may lead to an instability, when $\left(1 - n \frac{r_h'}{\gamma_h}\right) = 0$ which may be possible for a sufficiently large population. In this case, no steady state of PI (and SHI) is possible (see below). In the context of smart city, the case of the dominance of automata (a) is still more interesting, where (h) $\ll$ (a).

We assume $\tilde{r}_a \approx r_a$ and obtain

(44) $B = \gamma_h^{-1} n\ r_h (r_a a)^{-1}$

(45) $D = a^2 r_a r_a' \gamma_h (n\ \gamma_a r_h)^{-1}$

so that

(46) $BD = a\ r_a' / \gamma_a$
Putting (26), (28), (45), and (46) together, we obtain for the personal load

(47) $p_h/n = \left(1 - \left(a\ r_a'\right) / \gamma_a\right)^{-1} \left(\frac{n\, r_h}{a\, r_a} \cdot \frac{g_h}{r_h} + \frac{g_a}{r_a} \frac{r_h}{\gamma_h}\right)$

This is our most important result. As the first factor reveals, "smartification," i.e., increasing $a\ r_a' / \gamma_a$, *enhances* the personal load and may even lead to instability. Because of the composition of $a\ r_a' / \gamma_a$, this conclusion holds for both many small AI devices and for few large computer centers. The first term in the second bracket in (47) is proportional to the number n of citizens and describes an increase of personal load proportional to the size of the population. Note, however, that we consider the case $n\,r_h \ll a\,r_a$, i.e., strong smartification. The last term in the second bracket represents the cooperation of automata and humans. Since we don't ascribe "creative" power to the automata, g_a depends solely on the *import* of goods, or external ideas. If the import g_{hi}, g_{ai} as well as personal initiative g_{hc} tend to zero, a dying city will result.

6.1 The Information Crisis

The basic *time-dependent* equations read

(48) $\frac{dp_h}{dt} = n\,s\,r_h - \gamma_h p_h$

(49) $\frac{dp_a}{dt} = a\,s\,r_a - \gamma_a p_a$

(50) $\frac{ds}{dt} = n\,g_h + a\,g_a + n\,r_h' p_h + a\,r_a' p_a - \Gamma s$

where

(51) $\Gamma = \tilde{r}_h n + \tilde{r}_a a + \gamma_e$

To study stability, we consider the homogeneous Eqs., i.e., $g_h = g_a = 0$, and make the hypothesis

(52) $p_h = e^{\lambda t} p_{h0}\,, \quad p_a = e^{\lambda t} p_{a0}\,, \quad s = e^{\lambda t} s_0$

Inserting (52) into (48–50) leads to the eigenvalue equation

(53) $(\lambda + \Gamma)\,(\lambda + \gamma_h)\,(\lambda + \gamma_a) - n\,r_h'\,(\lambda + \gamma_a)\,n\,r_h - a\,r_a'\,(\lambda + \gamma_h)\,a\,r_a = 0$

For our purpose it suffices to consider the case where the automata are dominant

(54) $a^2\,r_a' r_a \gg n^2\,r_h' r_h$

Then (53) can be reduced to $\lambda = -\gamma_h$ or

(55) $(\lambda + \Gamma)\,(\lambda + \gamma_a) - a^2 r_a' r_a = 0$

The solution to (55) reads

(56) $\lambda_{+,-} = \frac{\Gamma + \gamma_a}{2} \pm \frac{1}{2}\left((\Gamma - \gamma_a)^2 + 4a^2 r_a' r_a\right)^{1/2}$

Instability leading to exponential growth occurs if $\lambda_+ > 0$. At the instability, $\lambda_+ = 0$, we find

(57) $\Gamma\gamma_a - a^2 r_a' r_a = 0, \quad \Gamma \approx \tilde{r}_a a$

which under the same approximations as above, leads to the previous instability condition

(58) $1 - \frac{a\,r_a'}{\gamma_a} = 0$

The exponential increase of SHI will lead to a communication breakdown so that this instability must be avoided. This can be achieved by technical innovations lowering the production rate of SHI, e.g., by reducing the transformation rate r_a' by many little improvements or a global breakthrough in AI. A "typical" approach will be increasing the costs of SHI – use (including taxation).

7 Final Notes

We conclude our paper with some notes on urban planning and design (UPD) in smart cities (SC): the common view is to see UPD as an external intervention in an otherwise spontaneous self-organized urban process. From this perspective, SC with their Internet of things (IOT), equipped with sensors covering the whole city, data mining techniques that enable to dig into, and exploit, big data, provide UPD stronger tools than ever to plan and design cities. Some would say, this transforms the current self-organized, somewhat chaotic, cities into centrally organized cities. And what about the citizens? The answer is that part of the SC machinery will be able to authentically identify and represent the citizens' views about the various plans and designs suggested by the professional planners/designers and implemented by the UPD authorities. This approach, as can be seen, assumes a fundamental distinction between the professional planners and designers and the planned citizens of the city. This is the common view, as noted, and big companies (e.g., IBM) see here a potential future market. Portugali's (2011) view is different: the lesson from the study of cities as self-organized systems is that every urban agent is a planner at a certain scale and that, in many cases, due to nonlinearities, the plan and design of a single person might be more dominant and influential than that of a whole team of professional planners. Examples are the stories of lofts, balconies, etc. (Portugali 2011). This view is further supported by cognitive science findings regarding humans' cognitive chronestetic capabilities for mental time travel and the implied phenomena of cognitive planning and prospective memory (Portugali 2016b). From this perspective, urban dynamics is seen as ongoing self-organized and organizing interaction between the city's many agents/planners each with its specific local-, mezzo- or global-scale plan. The challenge of SC with their IOT would be to develop means to foster this process. In fact, Christopher Alexander et al.'s (1977) *Pattern Language* and the *Self-Planned City* (Portugali 2011, Chap. 16) are steps toward this aim. There is an additional thought in particular brought forward by Haken. The ongoing development and use of AI/IT lead us to reconsider problems in mathematical terms, because quite clearly AI is based on mathematics. According to mathematics there are several classes of problems:

1. Under given conditions, a problem (e.g., in decision-making) cannot be solved at all. This is not a purely academic issue but may have consequences when AI is applied to real-life problems (possibly even to traffic steering). Thus we must not expect miracles from AI.
2. In many cases a problem possesses several solutions (cf., e.g., Sect. 4 Synergetics).

In sociological context this may lead to a "conflict" situation. Two simple examples may illustrate the case:

Traffic

(a) Left-hand drive
(b) Right-hand drive

Both regulations are possible, but exclude each other. This conflict can be solved only collectively via government decisions or direct votes.

Name giving to married people

(a) Regulated by state law
(b) Left to couples

In this case also, the "conflict" is solved collectively. This example may seem far-fetched, but in fact, conflict situations are ubiquitous in cities and can hardly be solved by automata.

3. In a number of cases, there exists a *unique* solution. In such a case, it doesn't matter, whether the solution is found by humans or automata.
4. In multicomponent/multi-agent/complex systems, the processes are of a mixed deterministic/stochastic nature. This requires the study of scenarios (realizations) by a combination of human imagination and insights (see above) and computing power.

A city may be considered as laboratory for innovations for a better quality of life (cf., e.g., Batty et al. 2012) in which the decisions relevant to human welfare must be left to the citizens. In our paper, aside from some general remarks, we focused our discussion on the role of AI and information dynamics in cities. We believe that innovations by AI will play an ever-increasing role, especially when dealing with the "information crises."

As the name indicates, AI directs attention to the role of artifacts, the production of which forms one of the basic capabilities of humans, that is, the production of objects that in one way or the other replaces the natural capabilities of humans by artificial ones. Thus, some of the early stone tools (e.g., flint knives) replaced the natural human teeth and fingernails as cutting devices. As noted in the introduction, the emergence of cities some 5,500 years ago was associated with the invention of writing – among the "smartest" inventions of society – which has partly replaced human memory: a person or society can write their story or thoughts on a stone (or papyrus or tablet or paper or computer) and need not keep it in memory. Already in antiquity this situation entailed a dilemma about the relations between the artifact and the natural human capability. In *Plato* there is an interesting dialogue between Socrates and Phaedrus in which Socrates expresses his concern about writing:

> In fact, it will introduce forgetfulness into the soul of those who learn it: they will not practice using their memory because they will put their trust in writing, which is external and depends on signs that belong to others, instead of trying to remember from the inside, completely on their own. You have not discovered a potion for remembering, but for reminding; you provide your students with the appearance of wisdom, not with its reality.

> Your invention will enable them to hear many things without being properly taught, and they will imagine that they have come to know much while for the most part they will know nothing. And they will be difficult to get along with, since they will merely appear to be wise instead of really being so. (Plato. c.399–347 BCE. "Phaedrus." Pp. 551–552 in *Compete Works*, edited by J. M. Cooper. Indianapolis IN: Hackett)

This concern about innovative artifacts accompanies society for many years and is relevant today with the smart artifacts and cities. Looking back at history, we can see that writing was not associated with the deterioration of memory: rather it enabled the externalization and thus the extension of memory – a new form of division of labor between the artificial and natural memory (see SIRN in this respect – Haken and Portugali 1996). As in division of labor in general, so in the case of writing, the challenge was to find a steady state that maximizes the relative advantage of the human memory and that of the artificial memory. The same applies here: the challenge facing smart cities is to identify a steady state that maximizes the relative advantage of the human sensorium and intelligence and that of the artificial ones.

References

Alexander, C., Ishikawa, S., Silvestein, M.: A Pattern Language. Oxford University Press, New York (1977)

Barabási, A.-L.: Network Science. Cambridge University Press, Cambridge (2016)

Batty, M., Axhausen, K.W., Gianotti, F., Pozdnoukhov, A., Bazzani, A., Wachowicz, M., Ouzounis, G., Portugali, Y.: Smart cities of the future. Eur. Phys. J. Spec. Top. **214**, 481–518 (2012). doi:10.1140/epjst/e2012-01703-3

Bauernhansl, T.: Arena 2036 – the fourth industrial revolution in the automotive industry. In: Glanz, C., et al. (eds.) 15th Stuttgart international symposium. Automotive and Engine Technology – Vol. 1. Wiesbaden, Springer Vieweg (2015)

Bettencourt, L.M.A., Lobo, J., Helbing, D., Kühnert, C., West, G.B.: Growth, innovation, scaling, and the pace of life in cities. Proc. Natl. Acad. Sci. U. S. A. **104**(17), 7301–7306 (2007). doi:10.1073/pnas.0610172104

Bonnefon, J.-F., Shariff, A., Rahwan, I.: The social dilemma of autonomous vehicles. Science. **352**(6293), 1573–1576 (2016)

Bretagnolle, A., Daudé, E., Pumain, D.: From Theory to Modelling: Urban Systems as Complex Systems. Cybergeo: European Journal of Geography. Dossiers 2005–2002, Article 335. 13th European Colloquium on Theoretical and Quantitative Geography, Lucca, 8–11 September 2003 (2006)

Feder-Levy, E., Blumenfeld-Lieberthal, E., Portugali, J.: The Well-Informed City. To be published (2016)

Floridi, L.: The Philosophy of Information. Oxford University Press, Oxford (2011)

Floridi, L.: Semantic Conceptions of Information. Stanford Encyclopedia of Philosophy, Stanford (2015)

Haken, H.: Synergetics. Introduction and Advanced Topics. Springer, Berlin (2004)

Haken, H., Portugali, J.: Synergetics, inter-representation networks and cognitive maps. In: Portugali, J. (ed.) The Construction of Cognitive Maps, pp. 45–67. Kluwer Academic Publishers, Dordrecht (1996)

Haken, H., Portugali, J.: Information Adaptation. The Interplay Between Shannon and Semantic Information in Cognition. Springer, Berlin (2015)

Haken, H., Portugali, J.: Information and self-organization. A unifying approach and applications. Entropy. **18**, 197 (2016)

Holling, C.S.: Resilience of ecosystems; local surprise and global change. In: Clark, W.C., Munn, R.E. (eds.) Sustainable Development of the Biosphere, pp. 292–317. Cambridge, Cambridge University Press (1986)

Le Cun, Y., Bengio, Y., Hinton, G.: Deep learning. Nature. **521**, 436 (2015)

Neuman, J.V., Morgenstern, O.: Theory of Games and Economic Behavior. Princeton University Press, Princeton (1944)

Nowak, M.A.: Evolutionary Dynamics. Belknap, Cambridge, MA (2006)

Portugali, J.: Selforganization and the City. Springer, Berlin (2000)

Portugali, J.: Complexity, Cognition and the City. Springer, Berlin (2011)

Portugali, J.: Complexity theories of cities: achievements, criticism and potentials. In: Portugali, J., Meyer, H., Stolk, E., Tan, E. (eds.) Complexity Theories of Cities Have Come of Age: An Overview with Implications to Urban Planning and Design, pp. 47–62. Springer, Berlin (2012)

Portugali, J.: Interview in Lisa Kremer: "What's the Buzz about Smart Cities?". Tel Aviv University (2016a)

Portugali, J.: What makes cities complex? In: Portugali, J., Stolk, E. (eds.) Complexity, Cognition Urban Planning and Design, pp. 3–19. Springer, Heidelberg (2016b)

Portugali, J., Stolk, E. (eds.): Complexity, Cognition Urban Planning and Design. Springer, Heidelberg (2016)

Roscia, M., Longo, M., Lazaroiu, G.C.: Smart City by Multi-Agent Systems. In: Proceedings of ICRERA 2013 IEEE International Conference on Renewable Energy Research and Applications, pp. 371–376. (2013)

Schwab, K.: The Fourth Industrial Revolution, Kindle edn. World Economic Forum, Switzerland (2016)

Shannon, C.E.: A mathematical theory of communication. Bell Syst. Tech. J. **27**(379–423), 623–656 (1948)

West, J.H., Brown, G.B.: The origin of allometric scaling laws in biology from genomes to ecosystems: towards a quantitative unifying theory of biological structure and organization. J. Exp. Biol. **208**, 1575–1592 (2005). doi:10.1242/jeb.01589

Wireless Protocols for Smart Cities

Gaurav Sarin

Abstract Rather than network of people, the Internet of Things (IoT) is network that connects things. It is an interconnection of uniquely identifiable embedded computing devices. It enables things to access data and communicate with one another. Humans will continue to use the Internet, but the future Internet will also be a pipeline for nonhuman devices – machine to machine (M2M) communication. When one connects any type of devices, tons of data will be generated. The promise of IoT is that more automatic and more intelligent services provided by interconnected smart devices will be prevalent with minimal amount of human interaction. Most of the things connected to IoT are actually simple devices that are referred to as smart devices. The devices become smart when joined together with other devices. The whole is greater than the sum of its parts, because everything is communicating with everything else in an intelligent and automated fashion. Any given device connects to other surrounding and relevant devices to share collected data. This creates what experts call ambient intelligence, which results when multiple devices act in unison to carry out everyday activities and tasks using the information and intelligence embedded into the network.

Abbreviations

6LoWPAN	IPv6 over Low-Power Wireless Personal Area Network
AES	Advanced Encryption Standard
BLE	Bluetooth Low Energy
CoAP	Constrained Application Protocol
CON	Confirmable
CoRE	Constrained RESTful Environments
CPU	Central processing unit
DTLS	Datagram Transport Layer Security
GFSK	Gaussian frequency shift keying

G. Sarin (✉)
Delhi School of Business, New Delhi, India
e-mail: Gaurav.sarin@dsb.edu.in

S.T. Rassia, P.M. Pardalos (eds.), *Smart City Networks*, Springer Optimization and Its Applications 125, DOI 10.1007/978-3-319-61313-0_6

HCI	Host controller interface
HTTP	Hypertext Transfer Protocol
IEEE	Institute of Electrical and Electronics Engineers
IETF	Internet Engineering Task Force
IoT	Internet of Things
IP	Internet Protocol
ISM	Industrial Scientific and Medical
JSON	JavaScript Object Notation
M2 M	Machine to machine
MAC	Media access control
MQTT	Message Queue Telemetry Protocol
MTU	Minimum transmission units
NON	Non-confirmable
PAN	Personal area network
PHY	Physical
REST	Representational State Transfer
RFC	Request for Change
RST	Reset
SIG	Special Interest Group
SSL	Secure Socket Layer
TCP	Transmission Control Protocol
TLS	Transport Layer Security
TSCH	Time Slotted Channel Hopping
UDP	User Datagram Protocol
URI	Universal Resource Identifier
XML	Extended Markup Language

IoT connects the individual devices to network and to each other. The network is also connected to software and services that analyze data collected by connected devices and use that data to make decisions and initiate actions from the same or other devices. The backbone for IoT is the network. To build IoT, experts envision three stages –

(a) Stage 1: Device Proliferation and Connection – More and more consumer devices are connectable, from fitness trackers, to television, to thermostats. These devices connect wirelessly to the Internet.
(b) Stage 2: Making Things Work Together – The data from one device is transmitted to a second device, which then makes some sort of decision and initiates a given operation. This stage is about automating simple tasks and programming the necessary devices to do that.
(c) Stage 3: Developing Intelligent Applications – To take advantage of vast amounts of data collected from IoT, we need applications that can act on larger, more complex, and often more obscure data sets.

In the world of IoT, each device no matter what its size will be assigned a distinct Internet Protocol (IP) address. Under the IPv4 addressing scheme, the large number of devices that need to connect will exceed available number of IP addresses. The way out is to use IPv6. This new protocol allows 2^{128} unique IP addresses and permits full-scale implementation of IoT. It expands the number of available IP addresses by moving from a 32-bit addressing to 128-bit addressing. In the IoT ecosystem, there are some wireless protocols that will be used to develop and implement smart cities across the globe. This chapter discusses five key wireless protocols in detail. The protocols are 6LoWPAN, ZigBee, BLE, CoAP, and MQTT. The key features, operations, security, packet formats, architecture, design principles, networking basics, and few differences of these protocols are explained in a clear and concise fashion.

1 IPv6 over Low-Power Wireless Personal Area Network

Many wireless technologies are used at a city level, but 6LoWPAN is the most prominent. This protocol is suitable for low-power devices, which transmit little amount of data. It is able to support large number of devices over metropolitan-wide area. A citywide 6LoWPAN network consists of several LoWPAN networks that manage smaller geographical regions. LoWPAN networks connect to main server and to the Internet using edge routers. Edge router manages network discovery for devices within LoWPAN and routes traffic into and out of LoWPAN. Each router houses smaller nodes that serve as hosts or routers. Nodes connect to devices and sensors. At network end, the city should have operating software that can monitor, analyze, and act on the collected data that is massive in size (Fig. 1).

1.1 Main Features of 6LoWPAN

- Supports 16 and 64-bit IEEE 802.15.4 addressing scheme
- Useful with low-power radios – BLE or 802.15.4
- Header compression of UDP header, IPv6 extension, and base header
- Auto-configuration of network
- IPv6 MTU, 1280 bytes; IEEE 802.15.4 frame size, 127 bytes
- Support for Internet Routing Protocol (Internet Engineering Task Force Routing Protocol for Low-Power and Lossy Networks)
- Broadcast, multicast, and unicast support
- Support for use of link layer mesh (e.g. IEEE 802.15.5)
- Security (Fig. 2)

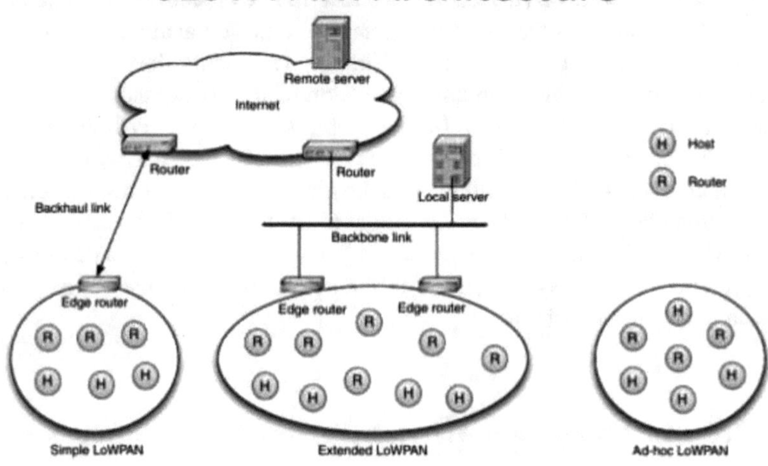

Fig. 1 6LoWPAN architecture (Source: The Internet of Things (Miller Pearson 2015))
Ad-hoc LoWPAN: No route outside the LoWPAN.
Simple LoWPAN: It has a single edge router.
Extended LoWPAN: Consists of multiple edge routers with common backbone link.
Edge Router: It runs special protocol, simplifies operation, and has a shared database

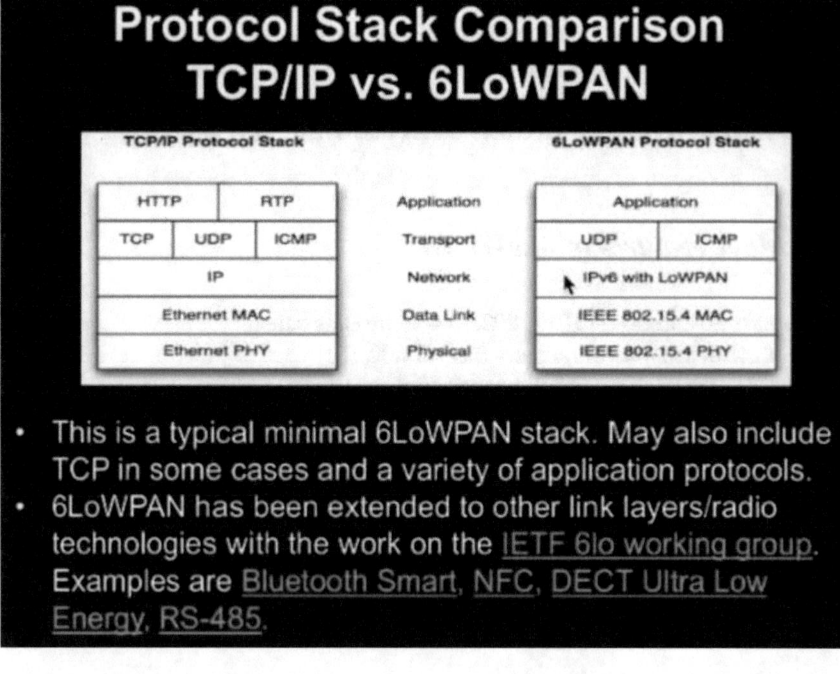

Fig. 2 TCP/IP vs 6LoWPAN (Source: https://www.edx.org/course/enabling-technologies-data-science-columbiax-ds103x-0)

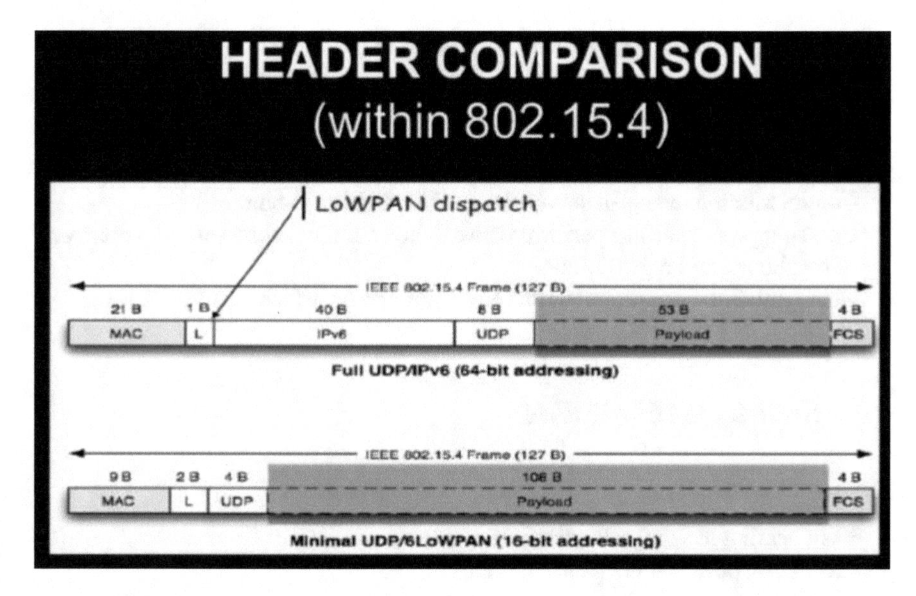

Fig. 3 Header comparison (Source: https://www.edx.org/course/enabling-technologies-data-science-columbiax-ds103x-0)

1.2 6LoWPAN Addressing

- IPv6 addresses are compressed in 6LoWPAN.
- LoWPAN has unique media access control addresses (Fig. 3).

1.3 6LowPAN Format Packet Fragmentation

Break a large packet into number of smaller ones.

- IPv6 has MTU of 1280 bytes.
- IEEE 802.15.4 leaves only 80–100 bytes of payload.
- RFC 4944 defines fragmentation and reassembly of IPv6.
- Poor performance of large IPv6 packets fragmented over wireless mesh networks:

  i. Lost fragments cause whole packets to be transmitted.
  ii. Low bandwidth and delay of wireless channel.
  iii. Application protocols should ideally avoid fragmentation by using compression or alternative protocols.

1.4 Setup and Operation of 6LoWPAN

Auto-configuration is a critical aspect for embedded networks.

Startup of 6LowPAN:

- Commissioning means to set up connectivity between nodes.
- Bootstrapping means to perform network layer address configuration, discovery of neighbors, and registrations.
- Route initialization means to run routing algorithm, which sets up the path.

1.5 Security for 6LoWPAN

In wireless networks, security is an important aspect since:

- Easy hearing of wireless radios is possible.
- Processing power of devices is limited.

The main security goals of a system:

- Confidentiality
- Integrity
- Availability

RFC 3552 specifies the threat model for internet security.

IEEE 802.15.4 has built-in encryption:

- Allows integrity check and encryption.
- AES-128 hardware engine is prevalent on most chips.

IP entails end to end security. RFC 4301 (IPsec standard) defines IP security.

2 ZigBee (Fig. 4)

Widely adopted in wireless networks, this protocol is low in cost and power. It is based on IEEE 802.15.4-2006 MAC and PHY standards. Current IEEE 802.15.4 physical layer suffices in terms of energy efficiency. Many IoT applications will exchange only few bits. The IEEE 802.15.4 MAC layer does not meet the requirements of different IoT applications. IEEE 802.15.4e standard modified the IEEE 802.15.4 MAC protocol. It talked about three MACs. TSCH mode facilitates energy-efficient communication, reducing fading and interference. IETF 6TiSCH working groups described Industrial IoT protocol stack that combined IETF 6LoWPAN, IEEE 802.15.4e TSCH, and IEEE 802.15.4 PHY layers. This protocol meets the requirements of high reliability, low latency, and ultralow jitter.

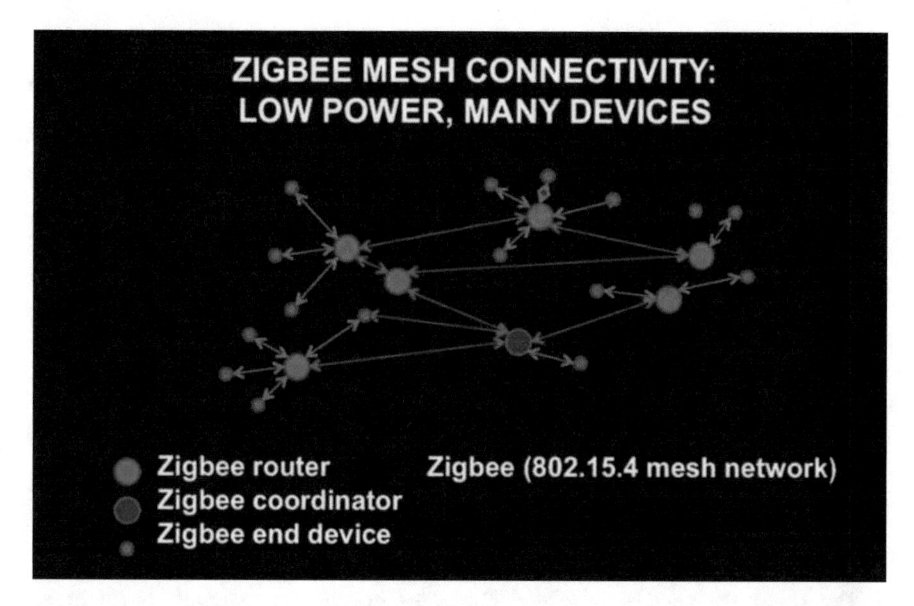

Fig. 4 ZigBee architecture (Source: https://www.edx.org/course/enabling-technologies-data-science-columbiax-ds103x-0)

2.1 Architecture Objectives

Ensure efficient use of available bandwidth and gives an implementation and platform for wireless networked devices:

- Ease to develop and design devices
- Reduce cost to build wireless solutions

  Provides interoperability using application profiles:

- Manufacturers can enable out of the box interoperable devices as desired.

  ZigBee models and network:

- Core functions and sensor device types
- Layers and modules
- Interfaces and services

2.2 Wireless Networking Basics

- Network Scan – Identify active channels in personal space
- Joining PAN – Create network on unused channel or join existing network in personal space

- Device Discovery – Identify device on active channels in PAN
- Service Discovery – Within PAN, find out the services that are available on devices
- Binding – Execute application level messaging

2.3 Wireless Networking Assumptions

Devices are pre-programmed to execute their desired network functions:

- To start a network, coordinator scans to figure out unused channel.
- To find an active channel, the router scans and then permits other devices.
- End device will definitely attempt to attach to already existing network.

To avail complementary services, device discovers other devices in the network:

- Any device in the network can initiate service discovery.
- Gateways are used to perform discovery outside the network but will be implemented in new ZigBee versions.
- For some special devices, binding provides command and control feature.

3 Bluetooth Low Energy

BLE is suitable for small range connectivity but suitable for low power, control, and monitoring applications. The protocol employs 2.4 GHz ISM band having 40 channels spaced 2 MHz apart. It has transmit data rate of 1 Mbps utilizing GFSK modulation. This protocol employs adaptive frequency hopping. To permit device discovery, it identifies 3 of the 40 channels to advertise. To transmit data, the remaining 37 channels are used. Hundreds of slave nodes can be connected to the master node. Range of BLE can be optimized according to application. Manufacturers can configure the range to 200 feet or more in-home applications. It supports single-hop technology, piconet, master device communicating with many slave nodes, and broadcast group technology, advertiser node broadcasting to several scanners. It will become key enabling technology for short distance IoT applications such as in healthcare, smart home, and smart energy domains. A broad market will reduce the costs of BLE hardware and will be used in cases where user-to-machine interactions are required (smart building, smart living, and smart healthcare). It is a modification of the traditional Bluetooth that cannot operate on a coin cell battery and has the following properties:

- A link is maintained when device is connected even if there is no data flowing. Hence it is connection oriented.

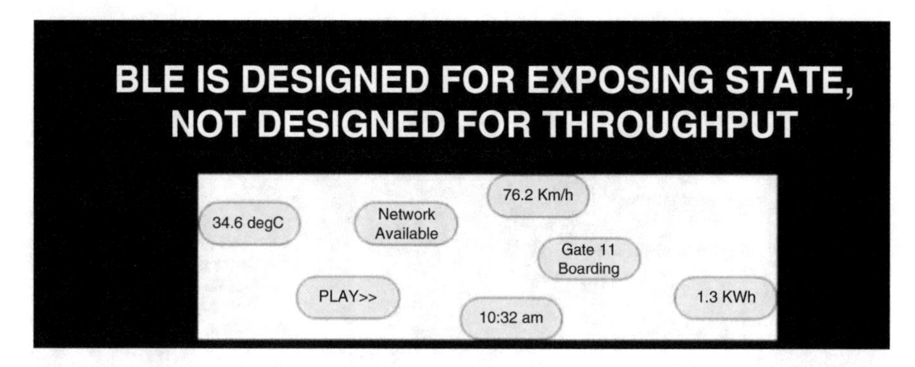

Fig. 5 BLE state example (Source: https://www.edx.org/course/enabling-technologies-data-science-columbiax-ds103x-0)

- To have battery life of several months, sniff modes make devices sleep, thereby reducing power consumption.
- Peak transmit current is approximately 25 mA. Although this is lower power than that required for other radio mediums, it is not low enough to be used in energy harvesting applications and coin cells.

  In BLE, everything is optimized for lowest possible energy consumption.

- Transmit peak current reduced via short packets.
- Receive time reduced via short packets.
- To improve discovery and connection time, less radio-frequency channels are used.
- State machine is simple (Fig. 5).
- BLE does not support streaming; hence data throughput is not a useful measure.
- It has the data transfer rate of 1 Mbps, but it is not optimized for file transfer.
- It is designed for exposing device state.
- Data can be triggered by local events.
- Data can be read by client at any time.
- Interface model is very simple.

3.1 BLE Protocol Stack

Controller and host are the two components of the stack (Fig. 6).

- Controller: System on chip with radio, link, and physical layers.
- Host has upper layer functionality and executes on application processor.
- Host/controller communication using HCI.
- Noncore profiles can run on top of the host.

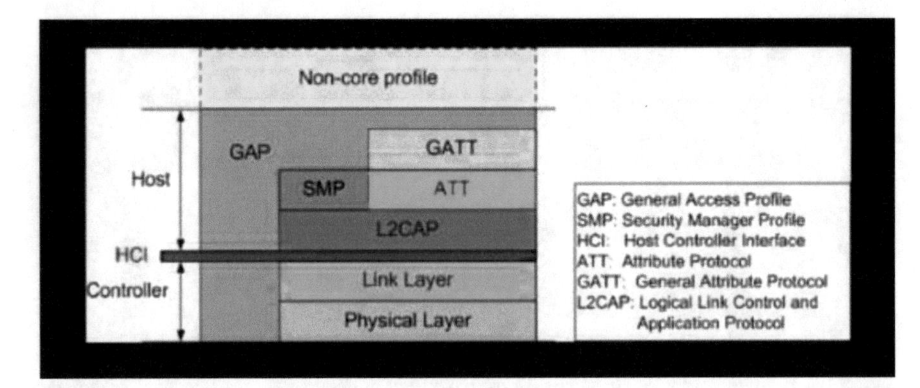

Fig. 6 BLE protocol stack (Source: https://www.edx.org/course/enabling-technologies-data-science-columbiax-ds103x-0)

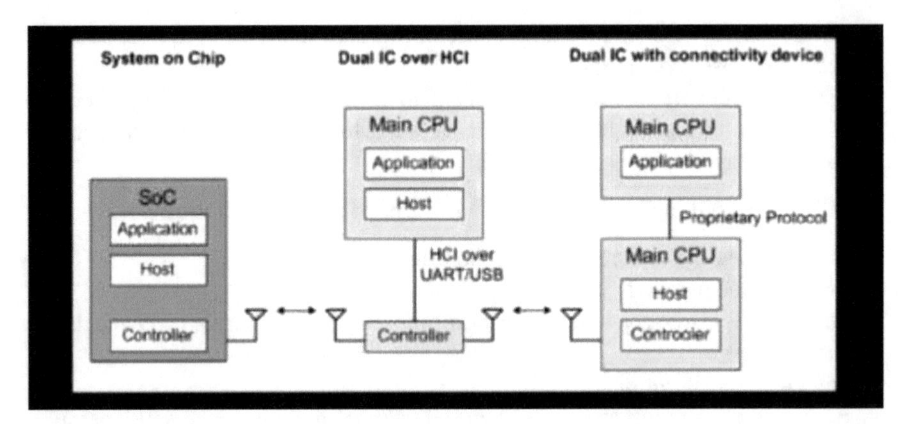

Fig. 7 BLE hardware configurations (Source: https://www.edx.org/course/enabling-technologies-data-science-columbiax-ds103x-0)

3.2 BLE Hardware Configurations

System on Chip – Used by simple sensors for low cost and complexity
Dual IC over HCI – Phones/tablets as they have powerful CPU
Dual IC with connectivity device – Additional microcontroller on which BLE connectivity is provided (Fig. 7)

3.3 BLE Advertising and Data Channels

Advertising channels – Useful for broadcast information, connection establishment, and device discovery. To minimize overlapping with IEEE 802.11 channels 1, 6, and 11, these channels are assigned center frequencies (Fig. 8).

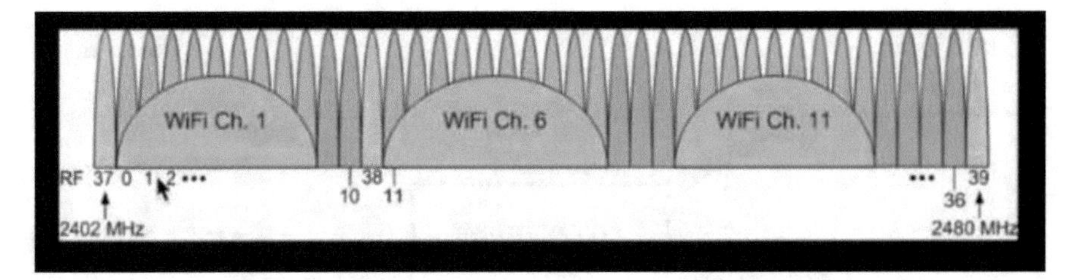

Fig. 8 BLE advertising channels (Source: https://www.edx.org/course/enabling-technologies-data-science-columbiax-ds103x-0)

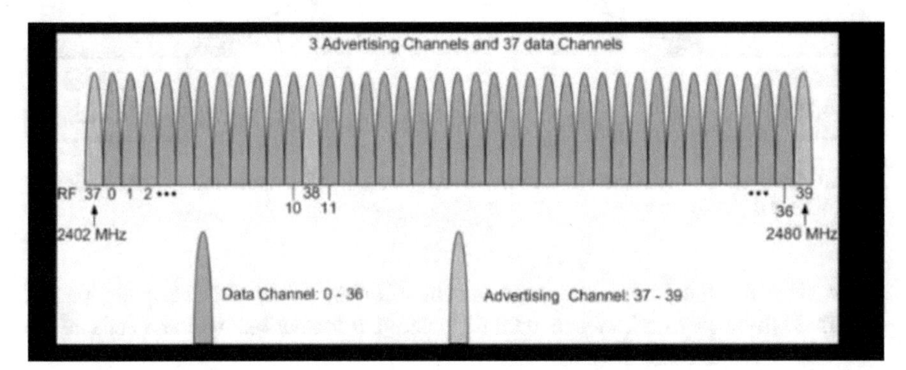

Fig. 9 BLE data channels (Source: https://www.edx.org/course/enabling-technologies-data-science-columbiax-ds103x-0)

Data channels between the connected devices are useful for bidirectional communication (Fig. 9).

3.4 BLE Broadcasting and Observing

Broadcaster – Sends advertising packets that are non-connectable periodically to anyone who wants to receive the same (Fig. 10).

Observer: To receive non-connectable advertising packets that are broadcasted, it repeatedly scans preset frequencies.

Central (Master) – For connectable advertising packets, central scans the predefined frequencies and when suitable initiates connection. Once connection is made, central will perform data exchange and will manage the timing.

Peripheral (Slave) – Periodically accepts connections that are incoming and sends advertising packets that are connectable. It abides the timing of central and exchanges data regularly with it in an active connection.

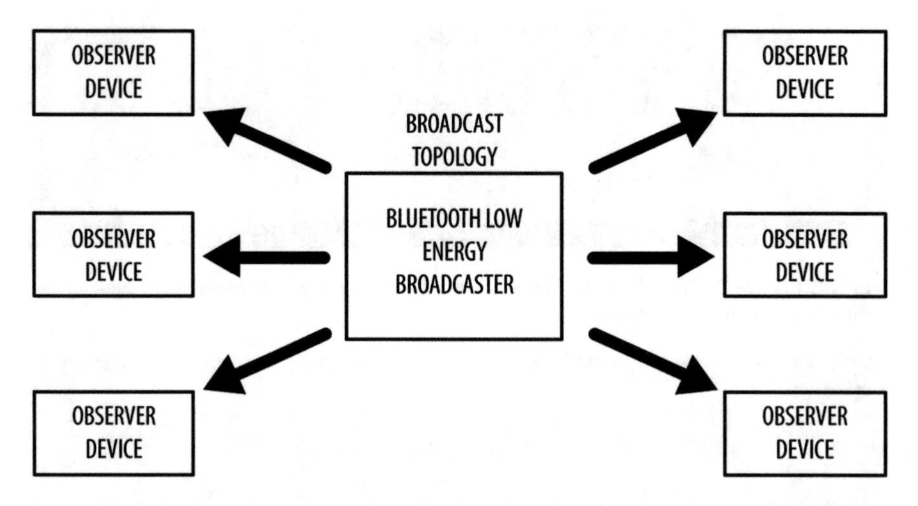

Fig. 10 BLE topology (Source: https://www.edx.org/course/enabling-technologies-data-science-columbiax-ds103x-0)

- To describe the broadcaster and its capabilities, standard advertising packet has 31-byte payload. It can include custom information, which needs to be broadcasted to other devices.
- BLE can provide optional secondary advertising 31-byte payload.
- Broadcasting is fast and easy, primarily used to push little amount of data to multiple devices or on a fixed schedule.
- Broadcasting has no security provisions, when compared to a BLE connection.

3.5 BLE Connections

One needs a connection if more than two advertising packets or in both directions data needs to be transmitted. Connection is periodical and a permanent data exchange of packets between two devices. Data is not sent to any other device except sent and received by two peers involved in a connection.

4 Constrained Application Protocol (Fig. 11)

The goal of CoRE working group was to bring for constrained nodes and networks, the REST architecture in a suitable form. A request response protocol, CoAP provides reliable and unreliable communication. URI is required for resource access on given host both in CoAP and HTTP. Device that is CoAP-enabled device can act

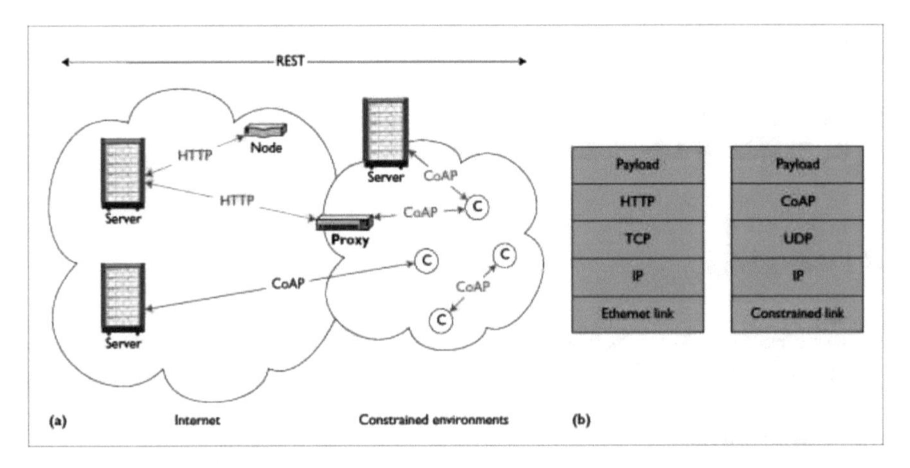

Fig. 11 CoAP architecture (Source: https://doi.ieeecomputersociety.org/cms/Computer.org/dl/mags/ic/2012/02/figures/mic20120200622.gif)

as server, client, or both. It can send non-confirmable messages. The protocol is defined for constrained networks, rather than reusing HTTP, because of reduction in code size and in bandwidth requirements. Moreover, this reduces latency and enhances reliability. It executes on the UDP layer and supports both multicast and unicast. HTTP over TCP supports unicast only. Four types of messages used by the protocol for message exchange between server and client:

(a) CON Messages – Carry request or response and require acknowledgment.
(b) NON Messages – Used for repeated messages and do not require acknowledgment.
(c) Acknowledgment – CON messages are acknowledged and must be empty or response.
(d) RST – Sent if some context is missing or CON message is not received properly.

CoAP is/has:

- RESTful protocol.
- Easy to proxy to/from HTTP.
- Suited for constrained devices and networks.
- Web protocol that is constrained and meets the M2M requirements.
- Asynchronous message exchanges.
- Simple proxy and caching capabilities.
- Low header overhead and parsing complexity.
- URI and content-type support.
- RFC0768 provides UDP binding and supports unicast and multicast requests.
- RFC6347 provides security binding to DTLS.
- Via HTTP, CoAP resources can be accessed in a uniform way.

CoAP is not:

- A general replacement of HTTP
- HTTP compression
- Restricted to isolated automation networks

4.1 Caching

- To decrease network bandwidth and response time, CoAP may cache responses.
- The primary goal of caching in CoAP is to reuse a prior response message to cater to the current request.
- Freshness model:
  (i) Without contacting the origin server, a response that is fresh in cache will be used for subsequent requests, hence improving overall response time.
  (ii) An origin server provides explicit expiration time by incorporating max-age option to estimate the freshness (Figs. 12 and 13).

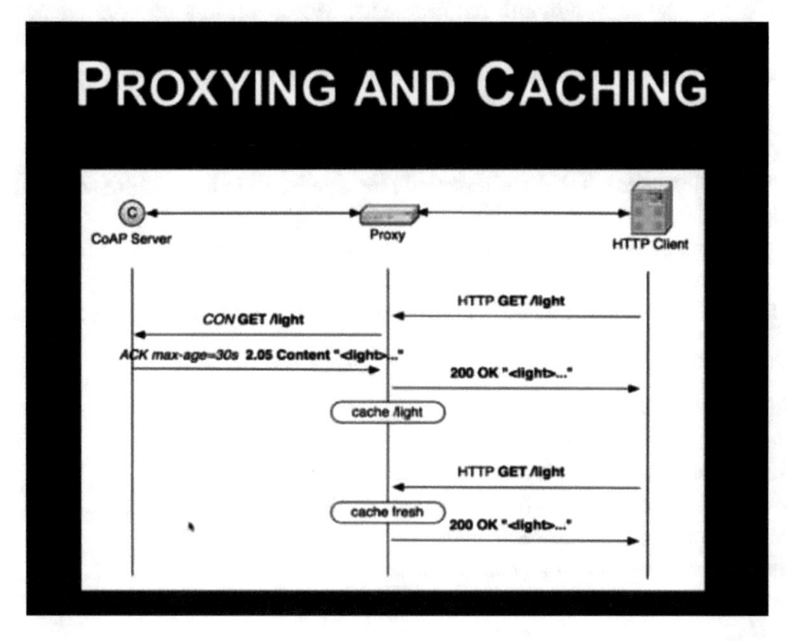

Fig. 12 CoAP proxying and caching (Source: https://www.edx.org/course/enabling-technologies-data-science-columbiax-ds103x-0)

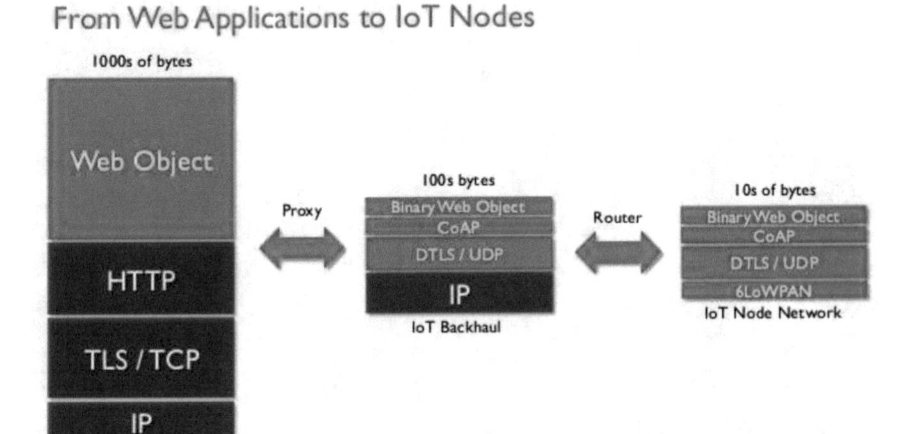

Fig. 13 CoAP from Web applications to IoT nodes (Source: http://image.slidesharecdn.com/coap-iot-tutorial-130519145022-phpapp02/95/arm-coap-tutorial-5-638.jpg?cb=1398894174)

5 Message Queue Telemetry Protocol

MQTT rely on message-oriented middleware. It focuses on delivering publish-subscribe messaging service via topics. It is suitable for IoT applications since it has small footprint and low bandwidth requirement. This protocol can be used for variety of environments – home automation application. It is easy to implement, is light, and has a low bandwidth system to support large number of devices. It has publisher, server, and subscribers. The protocol has user-password authentication feature. The protocol does not have any other authorization mechanism. The Amazon Web Services IoT ecosystem uses a MQTT broker.

We are in search of protocols for the constrained environments. How to cross firewall boundaries is a concern with protocols such as CoAP and HTTP (Fig. 14).

5.1 MQTT Goals

- Connectivity is given outside enterprise boundaries to the sensor devices.
- Multiple connectivity possible for remote devices and sensors.
- Gives proper data to decision-making and intelligent assets, which can leverage the same.
- Helps in deployment scalability and solutions management.

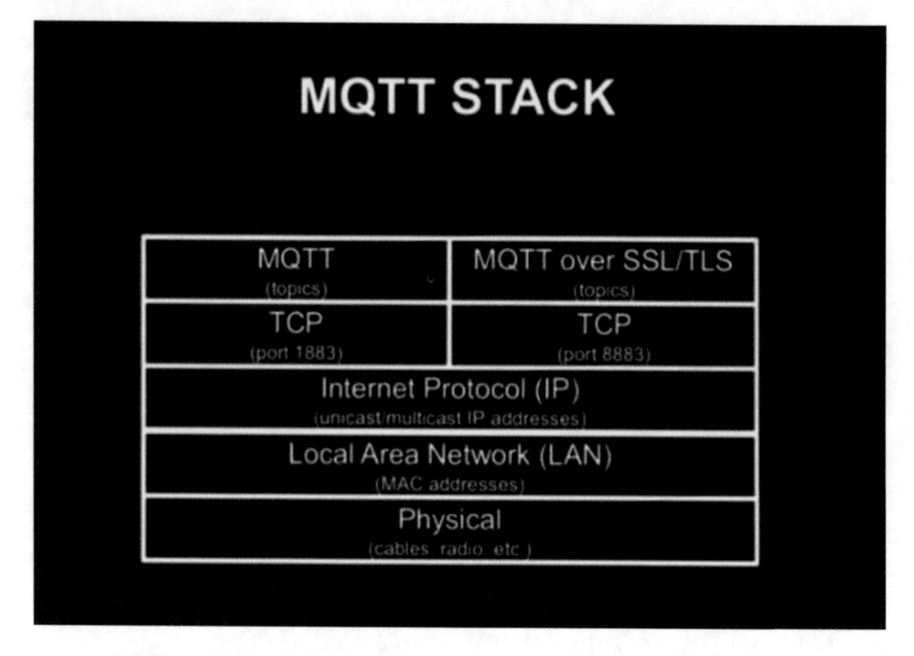

Fig. 14 MQTT Stack (Source: https://www.edx.org/course/enabling-technologies-data-science-columbiax-ds103x-0)

5.2 *MQTT Features*

- Extremely simple and lightweight messaging protocol
- Useful for constrained devices consuming low bandwidth having high latency or for unreliable networks
- Uses publish-subscribe methods
- Open source

5.3 *MQTT Design Principles*

- Minimize network bandwidth
- Minimize device resource requirements: memory and computation
- Ensure reliability
- Some degree of delivery assurance (Fig. 15)

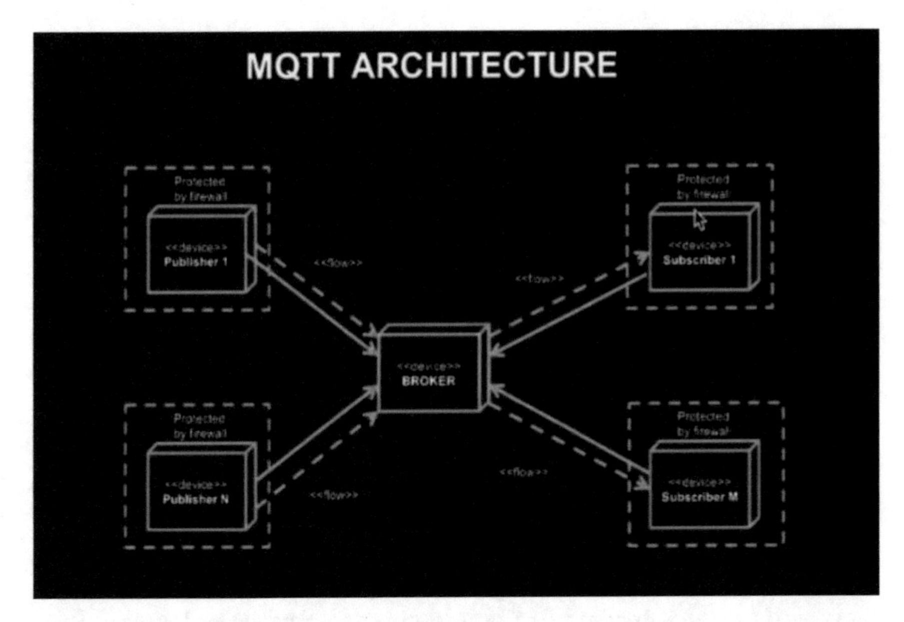

Fig. 15 MQTT architecture (Source: https://www.edx.org/course/enabling-technologies-data-science-columbiax-ds103x-0)

5.4 MQTT Publish and Subscribe

The MQTT protocol works on publish-subscribe pattern, as opposed to the request-response and pattern.

The publish-subscribe mechanism has three types of actors:

(a) Publisher: The role of the publisher is to connect to the message broker and publish content.
(b) Subscriber: Connect to the same message broker and subscribe to content that they are interested in.
(c) Message Broker: This makes sure that the published content is relayed to interested subscribers.

Principle to subscribe to topics and publish messages:

- Clients get messages from topics they subscribe to.
- To make messages available to all subscribers, clients can publish messages to topics.

When publishing content, the publisher can choose whether the content should be retained by the server. If retained, each subscriber will receive the latest published value directly when subscribing. Topics are ordered into a tree structure of topics, like a filesystem. The forward slash character (/) is used as a delimiter when describing a topic path. Subscription can be done by subscribing to either a specific topic by providing its path, or an entire branch using the hash wildcard character (#). There is also a single level wildcard character: the plus character (+).

Example: Sensor will publish measured temperature on the topic.

Clayster/LearningIoT/Sensor/Temperature

Clayster/+/Sensor/# will subscribe to all the sub-branches of the Sensor class that start with Clayster, and then any subtopic, which in turn will have a Sensor subtopic.

5.5 MQTT Levels of Service

The lowest level is an unacknowledged service. Here, the message is delivered at most once to each subscriber. The next level is an acknowledged service. Here, each recipient acknowledges the receipt of the published information. If no receipt is received, the information can be sent again. This makes sure the information is transmitted at least once. The highest level is called the assured service. Here, information is not only acknowledged but sent in two steps. The first is transmitted and then delivered. Each step is acknowledged. This helps to deliver content to each subscriber exactly once.

5.6 MQTT Messages

The server retains message after sending to all subscribers. The new subscribing client will get any retained messages if new subscription is sought for the same topic. A clean session flag is set, when MQTT client connects to the server:

When client disconnects from server, the subscriptions are removed if clean session flag has value true. If the clean session flag has value false, the subscriptions of the client are retained after connection termination, and the connection is durable. Further the messages that have high level of QoS (Quality of Service) are stored for subsequent delivery. The client may or may not set the clean session flag since it is optional.

5.7 MQTT Wills

- In case of unexpected disconnection, client informs the server of holding will or message, which will be published to relevant topic(s).
- Useful in security scenarios in which system managers should know as early as possible when a sensor device lost connection with network.

5.8 MQTT Security Considerations

The support for user authentication in MQTT is weak.

Plain text username and password authentication provide an obvious risk if the server is not hosted in a controlled environment. To circumvent this problem, MQTT can be used over an encrypted connection using SSL or TLS.

Other methods, not defined in the MQTT:

Use of client-side certificates or pre-shared keys to identify clients, instead of using the username and password option provided by the protocol. Proprietary methods of encrypting the contents to make sure only receivers with sufficient credentials can decrypt the contents (Fig. 16) (Table 1).

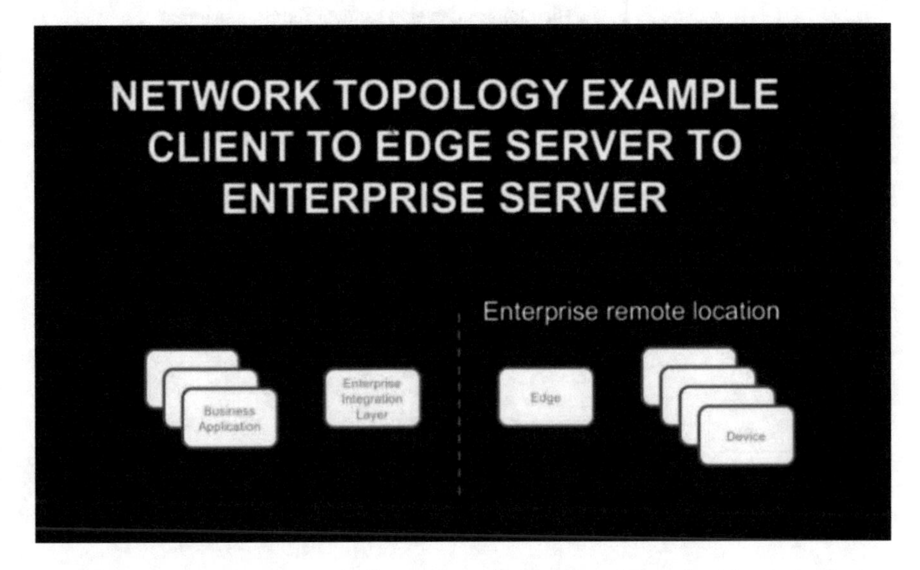

Fig. 16 Network topology example (Source: https://www.edx.org/course/enabling-technologies-data-science-columbiax-ds103x-0)

Table 1 Network topology description

Component	Purpose
Business applications	Application that sends and receives messages. Using transport protocol MQTT or JMS, connects to server in enterprise integration layer
Enterprise integration layer	Works as central concentrator to which remote devices connect. The server supports the following other activities:
	Service bus for business applications
	Message sharing
	If connectivity was not available, it is capable of storing notifications and events for near future transmittal
	Enforces connection security by using SSL/TLS to authenticate sensor devices
	To decrease the message volume flowing over network, it mediates to filter and aggregate messages
	Event/alarm mapping or merging between business applications and end user devices
Edge gateway	Works as hub on which devices attach. In addition, the following actions are supported:
	Gateway for enterprise integration layer
	Without enterprise integration layer, devices communicate with each other
	If connectivity was not available, it is capable of storing notifications and events for near future transmittal
	Apply application logic to aggregate or filter messages
	Apply application logic to map or merge event or alarm between enterprise integration layer and sensor device
Devices	Instrument that run tasks to communicate with central system or gather information

Source: https://www.edx.org/course/enabling-technologies-data-science-columbiax-ds103x-0

6 Few Comparisons (Tables 2, 3, and 4)

Table 2 Comparison between communications technologies

	NFC	RFID	Bluetooth®	Bluetooth® LE	ANT	Proprietary (Sub-GHz and 2.4 GHz)	Wi-Fi®	ZigBee®	Z-wave	KNX	Wireless HART	6LoWPAN	WiMAX	2.5–3.5 G
Network	PAN	PAN	PAN	PAN	PAN	LAN	LAN	LAN	LAN	LAN	LAN	LAN	MAN	WAN
Topology	P2P	P2P	Star	Star	P2p, star, tree, mesh	Star, mesh	Star	Mesh, star, tree	Mesh	Mesh, star, tree	Mesh, star	Mesh, star	Mesh	Mesh
Power	Very low	Very low	Low	Very low	Very low	Very low to low	Low–high	Very low	Very low	Very low	Very low	Very low	High	High
Speed	400 Kbs	400 Kbs	700 Kbs	1 Mbs	1 Mbs	250 Kbs	11–100 Mbs	250 Kbs	40 Kbs	1.2 Kbps	250 Kbs	250 Kbs	11–100 Mbs	1.8–7.2 Mbs
Range	<10 cm	<3 m	<3 m	5–10 m	1–30 m	10–70 m	4–20 m	10–300 m	30 m	800 m	200 m	800 m (Sub-GHz)	50 km	Cellular network
Application	Pay, get access, share, initiate service, easy setup	Item tracking	Network for data exchange, headset	Health and fitness	Sports and fitness	Point to point connectivity	Internet, multimedia	Sensor networks, building, and industrial automation	Residential lighting and automation	Building automation	Industrial sensing networks	Sensor networks, building, and industrial automation	Metro area broadband Internet connectivity	Cellular phones and telemetry
Cost adder	Low	Low	Low	Low	Low	Medium	Medium	Medium	Low	Medium	Medium	Medium	High	High

Source: Freescale and ARM White Paper

Table 3 Bluetooth vs BLE vs ZigBee

Specifications	Bluetooth	Bluetooth Low Energy	ZigBee
Standard	IEEE 802.15.1	Bluetooth SIG	IEEE 802.15.4
Range	100 m, optimized for 10 m	50 m, optimized for 10 m	75 m
Frequency	2.4 Giga Hz	2.4 Giga Hz	2.4 Giga Hz, 868 or 915 Mega Hz
Nodes	7	Not defined	65,635
Data rate	1–3 Mega bps	1 Mega bps	20–250 Kilo bps
Security	56/128 bit	128-bit AES	128-bit AES
Latency	100 ms	6 ms	30 ms
Power consumption	1 Watt	0.01 to 0.5 Watt	–
Peak current consumption	< 30 mA	< 15 mA	60 mW
Lifetime	Several weeks	Months or years	Months or years

Source: https://www.edx.org/course/enabling-technologies-data-science-columbiax-ds103x-0

Table 4 MQTT vs HTTP

	MQTT	HTTP
Design focus	Data	Document
Design pattern	Publish and subscribe	Request and response
Complexity	Low	High
Size of message	Small in size having two-byte header	Large in size as status is text
Level of service	Three service levels	The same level of service for all the messages
Extra libraries	Java and C libraries	Application dependent (JSON and XML)
Data distribution	Supports one to n, one to one, and one to zero	Supports one to one only

Source: https://www.edx.org/course/enabling-technologies-data-science-columbiax-ds103x-0

References

Asghar, M. H., & Mohammadzadeh, N.: Design and simulation of energy efficiency in node based on MQTT protocol in Internet of Things. In Green Computing and Internet of Things (ICGCIoT), 2015 International Conference on, pp. 1413–1417. IEEE (2015)

Collotta, M., Pau, G.: A novel energy management approach for smart homes using bluetooth low energy. IEEE Journal on Selected Areas in Communications. **33**(12), 2988–2996 (2015)

Farokhmanesh, F.: Analyzing and evaluating network protocols in IoT. In IEEE Ninth International Conference (2014)

Hersent, O. et al.: The internet of things: Key applications and protocols. Wiley (2015)

http://ocw.cs.pub.ro/courses/iot. Bucharest University

https://www.edx.org/course/enabling-technologies-data-science-columbiax-ds103x-0. Columbia University

Karimi, K., & Atkinson, G.: What Internet of Things needs to become a reality. EE Times July, 19 and Freescale, ARM White Paper (2015)

Keoh, S.L., Kumar, S.S., Tschofenig, H.: Securing the internet of things: a standardization perspective. IEEE Internet of Things Journal. **1**(3), 265–275 (2014)

Miller, M.: The Internet of things: how smart TVs, smart cars, smart homes, and smart cities are changing the world. Pearson Education (2015)

Modoff et al.: The Internet of Things, Deutsche Bank Markets Research (2014)

Niruntasukrat, A., Issariyapat, C., Pongpaibool, P., Meesublak, K., Aiumsupucgul, P., & Panya, A.: Authorization mechanism for mqtt-based internet of things. In Communications Workshops (ICC), 2016 IEEE International Conference on, pp. 290–295. IEEE (2016)

Tayur, V.M., Suchithra, R.: Internet of things architectures: modeling and implementation challenges. IJITR. 9–13 (2015)

Vermesan, O., & Friess, P. (eds.).: Internet of things: converging technologies for smart environments and integrated ecosystems. River Publishers (2013)

Webb, W.: The Role of Networking Standards in Building the Internet of Things (2012)

Zanella, A., Bui, N., Castellani, A., Vangelista, L., Zorzi, M.: Internet of things for smart cities. IEEE Internet of Things Journal. **1**(1), 22–32 (2014)

Leveraging Smart City Projects for Benefitting Citizens: The Role of ICTs

Renata Paola Dameri and Francesca Ricciardi

Abstract The scientific literature converges in indicating better life conditions for citizens as the smart city's main goal. To achieve this goal, cities leverage different technologies and especially ICT to modify urban infrastructures, public and private services and governance activities. However, smart programs often target the use and experimentation of innovative technologies, whilst citizens are considered as the passive addressees of technological programs. To verify whether smart projects really pursue citizens' well-being, an extensive empirical survey has been conducted. The research investigates 366 European smart city projects and extracts 42 ICT-enabled projects explicitly focusing on citizens. The analysis sheds light on the complex goal of citizens' well-being improvement in smart cities and on the most promising ICT solutions to impact urban life conditions. A special focus regards the use of IoT in smart projects addressing the citizens' well-being.

1 Smart City and ICT: Using Technologies to Improve the Citizens' Quality of Life

Smart city is an emerging topic on using hi-tech applications to create better life conditions in large cities. This trend has been evolving from 20 years ago ahead, and it is now spread all over the world (Dameri and Cocchia 2013). Amongst innovative technologies to apply in urban areas, ICT plays a pivotal role. ICT is already used in the first example of city digitalisation several years ago, and it continues to connote the modern city profile (Dameri 2016a).

Before speaking about smart city, the label digital city was already known in both academic research and urban policies, as well as in project implementation. Amsterdam digital city is the first example in the world, where a municipality assumes

R.P. Dameri (✉)
Department of Economics and Business Studies, University of Genova, Genova, Italy
e-mail: dameri@economia.unige.it

F. Ricciardi
Department of Business Administration, University of Verona, Verona, Italy
e-mail: francesca.ricciardi@univr.it

S.T. Rassia, P.M. Pardalos (eds.), *Smart City Networks*, Springer Optimization and Its Applications 125, DOI 10.1007/978-3-319-61313-0_7

ICT as the main driver of its strategy to create a better quality of life in a large city and to sustain the social development of the urban (Ishida 2002; Cocchia 2014).

A digital city is defined as a city using ICT to offer high-quality services to its citizens and to create a virtual community to support inclusive development and participated democracy (Schuler 2002). Several authors define similar concepts such as intelligent city, information city, wired city, and so on (Anthopoulos and Fitsilis 2010; Komninos 2006; Ergazakis et al. 2004). If we carefully examine these different (but similar each other) definitions, we can discover common contents and finally a sort of standard idea crossing the different concepts. The first common element is the digital infrastructure to support data collection, information and communication and knowledge sharing. The second common element is the connectivity of people with each other and with private and public bodies, supporting an easier communication and a participated democracy. The third aspect is the virtualisation of the city: a digital city is not only the digital infrastructure but also the virtual representation of artefacts, people, institutions and events, creating a second urban life aside from the real one.

The digital component of a city is a fundamental part of a smart city: even if authors do not agree about the relationship between smart city and digital city, it is evident from both the scientific literature and the empirical environment that smart city and digital city are somehow two faces of the same coin (Cocchia 2014; Dameri and Sabroux 2014). Indeed, if we examine the most cited smart city definitions, it emerges that almost all of them include ICT as a crucial driver for a smart city implementation.

However, smart city is characterised by a larger scope with respect to digital city. In digital city, to digitalise the urban life is a goal for itself, and digitalisation is conceived like a source of benefits. In smart city, technology is not a goal but a means to reach other results and especially a better quality of life in the urban area. This better quality of life could be defined in very different ways. Giffinger et al. (2007) refer to the enhancement of citizens' features such as awareness, independency and participation. The California Institute (2001) outlines the transformation of life and work, thanks to ICT. In Caragliu et al. (2011) the focus is on sustainable economic growth and high quality of life, through participatory governance but also natural resource preservation. Dameri (2013) synthesises the smart city goals in benefit creation for citizens in terms of well-being, inclusion and participation.

It emerges that, thanks to the smart city, ICT is used like a critical resource to transform the urban life and generate benefits for citizens, influencing the environmental, political and economic conditions of their city. This general framework should be implemented when realising each smart project composing a larger smart city strategy. Each project has its own contents, actions and goals: but the citizens' well-being should be the ever-present aim of every smart project. The quality of life in a city should emerge like a large puzzle composed of the contribution of each smart project, each of them participating in a specific way to compose the final patchwork (Neirotti et al. 2014).

To better understand how ICT in smart city can improve the quality of life, we introduce an empirical analysis, examining a set of 42 projects out of a larger set of 366. This set of projects contains commitments submitted by European smart cities into the European Innovation Partnership on Smart Cities and Communities (see https://eu-smartcities.eu/). Selected projects have both the following features: they are focused on citizens, and their main technological sector is ICT. The investigation permits to analyse in details several aspects characterising these projects, such as the nature of the stakeholders, the geographical distribution of the projects in Europe, the pursued goals and the project impact on different perspectives of the quality of life. It emerges a heterogeneous picture, where ICT is really capable to affect the people life in several ways.

2 Smart City Goals: The Impact on Citizens' Well-Being and Quality of Life

As already said in the first section, smart city is an urban strategy aiming at creating better life conditions in a city, using high technologies, but also relying on the active participation of smart citizens in its implementation. The first and more cited smart city framework, suggested by Giffinger et al. (2007), includes six smart dimensions (Fig. 1), and smart people is one of them. The smartness is realised both through material instruments (such as mobility infrastructures or environment preservation) and through people behaviours in living, working, studying and so on.

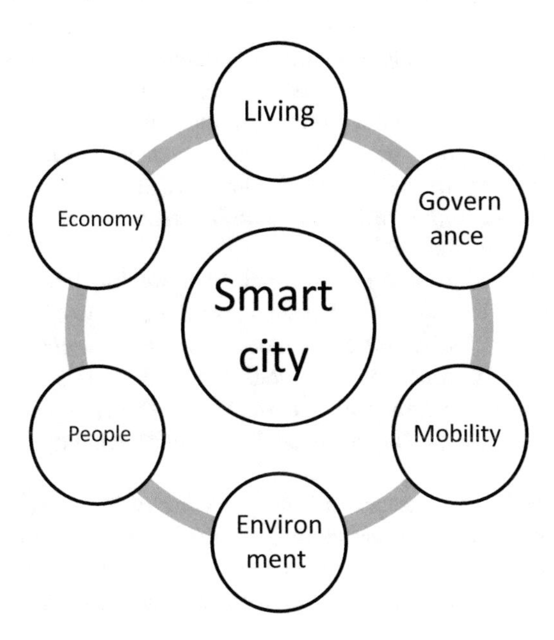

Fig. 1 Smart city dimensions (Giffinger et al. 2007)

Caragliu et al. (2011) include the people dimension in a smart city referring to human and social capital, outlining the importance of both the individual behaviours and the social relationships. Dameri (2013) suggests a smart city framework where the geographical and material dimensions of a smart city – territory and infrastructures – are a support for people and government to realise better life conditions in cities. In these and other definitions, people are not passive addressee of a smart city strategy but active players of the urban transformation: only by this path it is possible to obtain real improvements for citizens, thanks to smart technologies and infrastructures implemented in smart programs (Hollands 2008).

However, all these definitions neglect to say what a better life is, what a well-being condition is for people living in cities and how it is possible to measure or evaluate it and the smart city contribution to its increasing (Ricciardi and De Marco 2012). This aspect is central in the debate about smart cities and their goals, as the progressive urbanisation is more and more a source of both opportunities and diseases. Large cities are, all over the world, places where things happen: where people live, study and work and where the economic and scientific development is higher and the cultural offer (theatres, cinemas, exhibitions, etc.) the best. But cities are also the place where pollution, traffic, poverty and exclusion are the worst.

To support the analysis about the capability of smart city to affect the citizens' well-being, it is possible to refer to one of the most important well-being frameworks, the one introduced by OECD (2011). This framework is developed at both country level and local level (regions and metropolitan areas).

The OECD program is called Better Life Initiative and aims at defining and measuring well-being in all the OECD countries. Thanks to this framework, it is possible to understand what well-being is, define it, thanks to several concrete dimensions, but also to measure it, as each dimension is associated to one or more indicators. It is generally recognised that well-being depends not only on material life conditions, such as richness or health, but also on other life conditions such as environmental quality, social relationships, job richness and so on. Finally, the quality of life is a personal perception of each person deriving from needs and opportunities, how they are perceived and how are really met.

OECD suggests a comprehensive and multidimensional framework to define and measure well-being, as showed in Fig. 2.

It includes material conditions – income and wealth, job and earnings and housing – and other (less tangible) dimensions of the quality of life, such as health, education, civic engagement and environmental quality. Moreover, the framework can be applied in a dynamic way, as it considers both the present well-being and its sustainability in the medium-long term. Sustainability is based on stocks of human, social, natural and economic capital, on which smart city strategies are able to influence.

OECD applies the framework at both the country level and the local level, as evidence shows that some factors that most influence peoples' well-being are local issues. In this initiative, each region is measured in 11 topics – income, jobs, housing, health, access to services, environment, education,safety, civic engagement

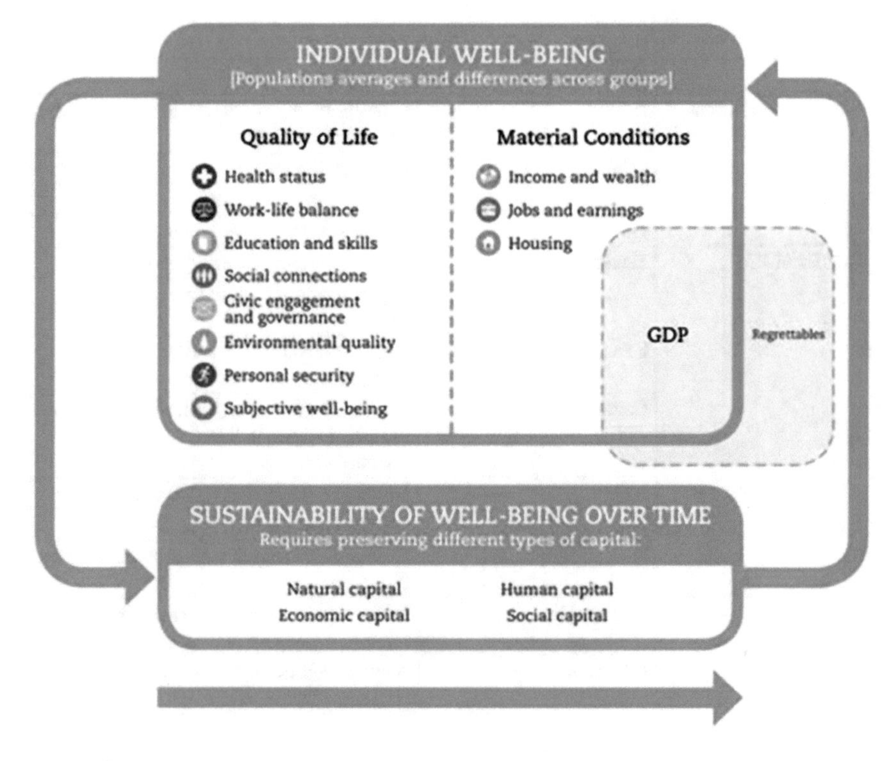

Fig. 2 OECD framework for measuring well-being and progress

and governance, community and life satisfaction. A score has been calculated for each topic so that you can compare places and topics within and across countries. Indicators are listed in Table 1.

In this work, the OECD framework is used to analyse the impact of smart city projects on well-being topics, as listed in Table 1. It permits to face the problems to understand and measure how smart city affects the daily life from a multidimensional point of view. The analysis is not quantitative but qualitative and permits to design a perimeter of the influence of smart initiatives using ICT on the quality of life in a city (Ricciardi and Za 2014). The OECD framework is therefore fundamental to give substance to the general assertion that smart city improves the citizens' well-being. Moreover, it permits to subdivide the global goal of quality of life in its components as suggested by OECD and to understand if smart city affects some topics more than the other and which they are.

Table 1 Well-being topics

	Topics	Indicators
Material conditions	Income	Household disposable income per capita (in real USD PPP)
	Jobs	Employment rate (%)
		Unemployment rate (%)
	Housing	Number of rooms per person (ratio)
Quality of life	Health	Life expectancy at birth (years)
		Age adjusted mortality rate (per 1000 people)
	Education	Share of labour force with at least secondary education (%)
	Environment	Estimated average exposure to air pollution in PM2.5 ($\mu g/m^3$), based on satellite imagery data
	Safety	Homicide rate (per 100,000 people)
	Civic engagement	Voter turnout (%)
	Accessibility of services	Share of households with broadband access (%)
Subjective well-being	Community	Percentage of people who have friends or relatives to rely on in case of need
	Life satisfaction	Average self-evaluation of life satisfaction on a scale from 0 to 10

3 An Empirical Analysis

The relationship between well-being in a city, smart projects and ICT is theoretically explained, but rarely it has been empirically verified. Moreover, as seen above, this concept is multi-faceted: well-being includes several topics; ICT is a bunch of very different solutions; smart projects are heterogeneous and city specific; citizens include different categories of people such as workers, students, tourists, elders and so on (Dameri 2016b). A deep and detailed analysis permits to better understand how and where smart projects are able to produce benefits, thanks to the use of ICT and who especially enjoys these benefits (Dameri 2012).

Our empirical analysis aims at investigating how real smart city projects use ICT to directly affect the citizens' life and which aspects of their life are influenced. The basis of our research is a large database of smart commitments, collected by the European Innovation Partnership on Smart Cities and Communities (EIP-SCC), an initiative supported by the European Commission bringing together cities, industry, SMEs, banks, research and other smart city actors. Its main goal is to improve the citizens' quality of life, support networking and partnering, accumulate knowledge and facilitate exchange of information about smart city solution implementation.

Commitments are a set of 366 projects involving more than 4000 partners from 31 countries. They include measurable and concrete smart city engagements and actions from public and private partners. Commitments are organised in six action clusters, covering all the aspects of a smart city, defined as follows:

- Business models, finance and procurement
- Citizen focus
- Integrated infrastructures and processes (including open data)
- Policy and regulations/integrated planning
- Sustainable districts and built environment
- Sustainable urban mobility

The citizen focus cluster is about industries, civil society and different layers of government working together with citizens to realise public interests at the intersection of ICT, mobility and energy in an urban environment. It especially collects projects addressing an enabling environment for citizens to solve their problems or projects that facilitate a conversation between stakeholders, where citizens' voices are not only heard but instrumental in solution design. Sixty one out of 366 commitments belong to this cluster. It is only 19%. From this first observation, it emerges that 4/5 of commitments are implemented without involving citizens as active partners.

Moreover, the commitment database permits to classify each entry depending on its main technological sector; these sectors are energy, transport, ICT or others. Filtering the 61 commitments in the citizen focus cluster with respect to the ICT as the main sector, we extract 42 projects. It means that 69% of commitments focusing on citizens use ICT as the main technology to reach the pursued results. ICT emerges like the most important driver when smart projects want to involve citizens actively. This selection is the basis of our empirical survey.

3.1 Research Methodology

Our research activity adopts both a quantitative analysis and a content analysis to extract knowledge from the data sheet describing the profile of each selected project. The quantitative analysis uses the tags assigned to each project by the author of the data sheet; the tag we consider for our analysis regards the geographical localisation of the project, identifying the European country where it is placed. Some projects regard more than one country as joint initiatives of more than one city in different EU members. For this reason, in the final report the total issues about the geographical distribution of smart projects are more than 42.

The manual content analysis permits us to add criteria that are not considered in the commitment database. We add three dimensions of analysis:

- Smart city content, classifying projects with respect to the six smart city dimensions suggested by Giffinger et al. (see Fig. 1)
- Addressees, considering which categories of citizens are more affected by the project
- Impact, classifying the projects with respect to the better life topics suggested by OECD (see Table 1)

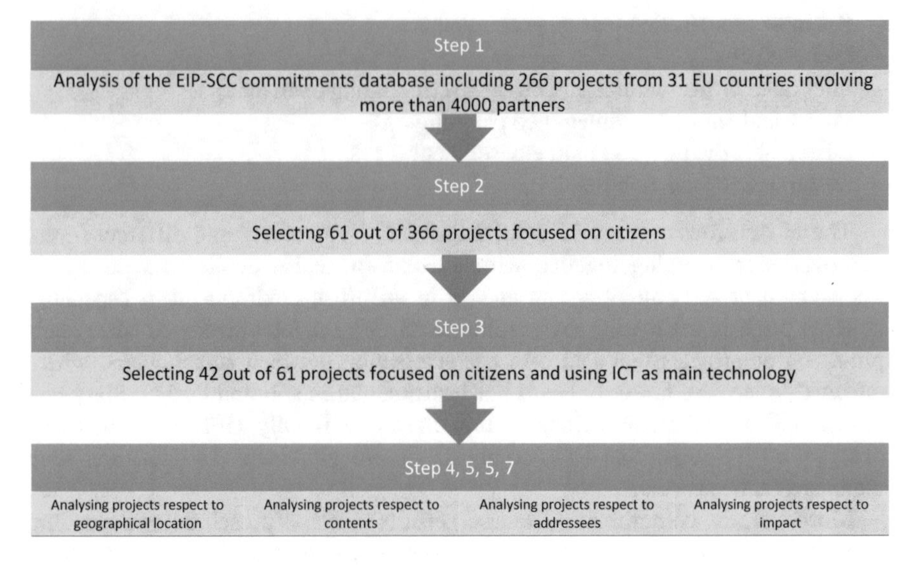

Fig. 3 Smart and digital project taxonomy

Summarising, our analysis evolves in this way (see Fig. 3).

Step 1: identification of the project database ➔ 366 projects from the EIP-SCC commitment database
Step 2: selection of projects with priority focused on citizens' well-being ➔ 61 out of 366 projects
Step 3: selection of projects focused on citizens' well-being and using ICT as the main technology ➔ 42 out of 61 projects
Step 4: classification of projects with respect to the country or countries where they are implemented
Step 5: classification of projects with respect to the six dimensions of Giffinger's smart city model
Step 6: classification of projects with respect to the addressees
Step 7: classification of projects with respect to the 11 well-being topics of the OECD Better Life framework

All these criteria are assumed by the commitment database or by well-known theoretical models (Giffinger smart city model and OECD Better Life framework). We defined keywords extracted from these theoretical models to examine the project descriptions and to tag them accordingly. With respect to step 6, we have not found a theoretical model classifying the different addressees of smart initiatives in smart cities. Therefore, we applied a two-step content analysis: in the first phase, we outlined all the keywords identifying one category of addressees found in the project data sheet, and counted the most recurrent ones. They are:

- Public bodies
- Entrepreneurs and firms
- Tourists
- People with disabilities
- Educators

In the second step, we tagged the projects accordingly with these more recurrent addressees.

Along with the formal content analysis, to read all the project data sheets permitted to us to explore the large and heterogeneous smart city project portfolio collected by EIP-SCC, along with the formal classification, we can suggest some general considerations, exposed in the last section of this chapter.

3.2 Results

The following results regard the 42 selected projects, focused on citizens and using ICT, as the main sector.

The geographical distribution of these projects in the 31 EU and non-EU[1] countries hosting the projects is very heterogeneous. As already said, several projects are implemented jointly by more than one city, in different countries; therefore, the total number of issues is higher than 42.

Examining the number of projects implemented in each country, several countries host no projects; on the contrary, some of them host numerous projects. It emerges therefore that the smart city phenomenon, and especially projects focused on citizens and using ICT, is not equally spread in Europe, even if EU Commission allocates large funds to this urban policy. Italy and Spain have the highest number of projects (Dameri 2016a); other countries have 7, 5, 3 and 2 1 projects each. This situation confirms other surveys regarding the distribution of smart city programs in Europe (EU Parliament 2014). Italy and Spain are more involved in this urban strategy, and their projects are spread in both large and medium cities, all over the national territory. The Netherlands is recognised as the pioneer country in Europe (and in the world) for smart strategies, and projects are mainly concentrated in Amsterdam, the first smart city.

Figure 4 shows the distribution of projects with respect to Giffinger's smart city dimensions. Also in this case, depending on the content analysis and the keywords found in the text describing the smart project, each project could be considered in more than one dimension; therefore, the total number in the histogram is more than 42.

[1]Some projects are funded by EU programs regarding not only EU members but also other countries with which EU has signed cooperation agreement.

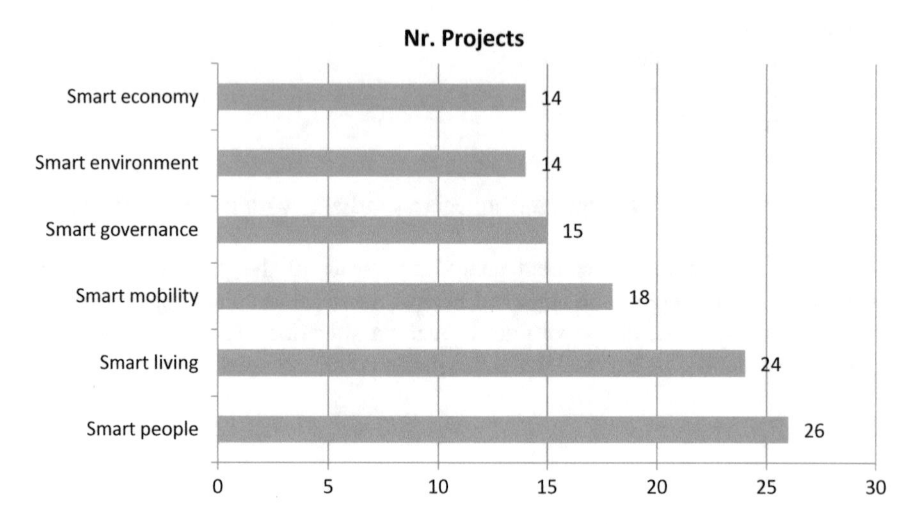

Fig. 4 Distribution of projects with respect to Giffinger's smart city dimensions

Twenty-six projects focus on smart people; this dimension regards the development of both human and social capital in smart city. Keywords identify aspects such as people education, professional skills and digital literacy and also social relationships and participation to the city governance and the political debate. These projects use ICT instruments such as open platform and open data, multimedia interfaces and knowledge sharing.

Twenty-four projects interest the smart living dimension; it regards the general well-being represented by the quality of housing, health services, security, education and touristic facilities. For example, several cities in Spain adopted an app based on a complex digital solution able to deliver information about the quality of beaches, public services, touristic infrastructures and so on. Rome offers a platform to improve the quality of life with respect to the relationship between citizens and the public administration, supporting digital activities in four different fields: entrepreneurship, event management, city security and tourism.

Eighteen projects involve the smart mobility dimension. It emerges also investigating the commitment database: 20 projects out of 42 are concurrently focused on both ICT and mobility as main sectors. In this group, we classify projects that use ICT to improve the citizens' quality of life improving the mobility facilities in the urban area. Projects especially regard the implementation of apps able to deliver real-time information about traffic, accidents, circulation of public transport or parking availability. Projects regarding mobility focused on citizens using ICT to communicate instead of using ICT to automate as it happens in other smart mobility projects where the citizens are not the main focus.

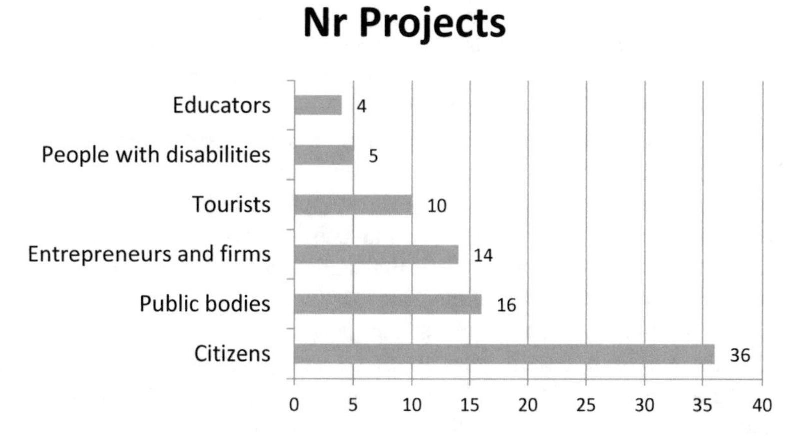

Fig. 5 Distribution of projects with respect to the addressees

Less projects are classified with respect to the other dimensions – environment, economy and governance. It is surprising that the smart governance dimension contains only 15 projects: perhaps local governments are not ready to implement smart facilities to better interact with their citizens.

Figure 5 shows the type of addressees most affected by the smart projects. Thirty six projects are addressed to all the citizens, without a special focus on any category.

All these projects are especially addressed to improve the citizens' quality of life by enhancing their participation to the political, cultural and social life of their city. Several projects are located in a specific neighbourhood, especially to face social problems in suburbs. Certain projects join the citizens' focus with the economic aspect of their life, offering platforms to collect business ideas and to facilitate the commercial development of enterprises, especially in disadvantaged urban areas. Generally, these projects have also the aim to test a pilot and to further replicate the best practices in other neighbourhoods.

Sixteen projects are addressed to the public bodies, as intermediaries towards the citizens. Municipalities are the main actors in planning, implementing and governing smart city large initiatives, but they need to invest in adequate infrastructures and facilities to support a better involvement of citizens in their governance. These projects aim at developing platforms in public administration to facilitate its relationship with citizens, to reduce bureaucracy and to increase transparency and accountability of public managers and officials. In some situations (e.g. in Italy), cities develop a joint project to share cost and multiply benefits in several urban areas.

Fourteen projects are addressed to entrepreneurs and firms, to enhance their capability to boost business. One mainstream of smartness for this category of addressee is to use ICT to communicate with the main stakeholders; we find two similar commitments (in Belgium and in the UK) developing a digital platform for delivering mentoring from expert managers to young entrepreneurs in innovative start-ups. In Portugal ICT is used to implement a digital showroom where local producers can promote their business ideas, products and services. By this type of commitments, a smart city aims at attracting newcos, support young entrepreneurship and facilitate business innovation.

Ten projects are devoted to tourists, considered as an important driver of business and development for cities, even the ones not traditionally known as touristic destinations. Especially municipalities develop apps and digital platforms to inform tourists about services, cultural events and exhibitions. In the meantime, these platforms are useful also for inhabitants that sometimes discover – thanks to smart platforms – the cultural identity of their own city.

Finally, smart commitments address also people with disability and educators. The first category can benefit from smart commitments in two ways. The first regards the use of ICT to inform about specific services delivered in the urban area to respond to disabled people-specific needs. The second way regards the use of ICT to overcome limits that these categories meet in their daily life.

Educators are addressees of four commitments aiming at supporting e-learning or knowledge sharing in schools, universities and also amongst the citizens. A special focus is the teaching method supported by these smart commitments: it emerges the aim to overcome traditional teaching activities to suggest open data, multimodality, augmented reality and so on.

Figure 6 shows the distribution of commitments with respect to their impact on the better life topics. Also in this clustering, each project can impact on more than one topic; therefore, the total issues are higher than 42.

It clearly emerges that life satisfaction is greatly the most important aspect pursued by 33 smart projects. It is not surprising, as the citizen focus is described like the active involvement of people in solving their own problems of daily life in

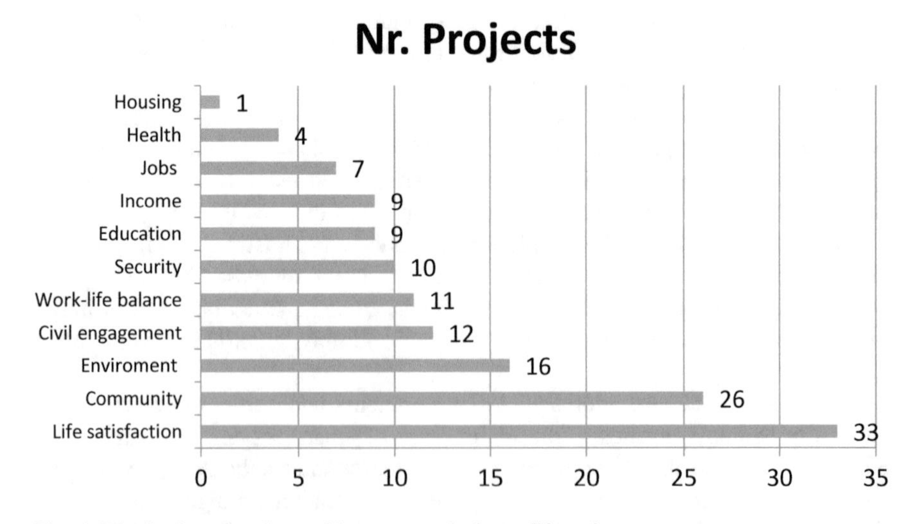

Fig. 6 Distribution of projects with respect to the better life topics

a city: it is implicit in the general aim of these commitments to satisfy each person, thanks to the ICT. Deeping the analysis of the commitments, these 33 smart projects can be further classified with respect to the ICT solutions implemented to meet the citizens' expectations.

Several commitments included in this cluster regard the implementation of platforms to realise an integrated smart city program, collecting more than one project regarding different aspects of the daily life but ever putting the citizens at the core of the program.

Other commitments address the sustainability of the urban life, pursuing the citizens' satisfaction together with the quality of the urban environment (including both the natural resources and the buildings or artefacts composing the urban fabric).

Another subset of projects regards the design and test of governance framework to involve citizens in the city regulation and planning; ICT is used like instrument to share opinions, ideas and points of view. Living labs are often the method to realise these projects.

A lot of projects included into this subset plan the development of mobile apps especially using geographic information systems and georeferencing to offer to citizens real-time and up-to-date information about a large bunch of urban services.

More generally, we can say that in this subset of 33 projects, we can find initiatives that regard a comprehensive view of smarter cities especially considering the active role of citizens not only in enjoying the benefits or in suggesting what to do but in realising the projects' aim, participating to the project implementation. These projects are more interesting in relation with the multidimensional nature of a smart city and its governance aspect. Also for this reasons, in this subset, we find a very high heterogeneity in project features, as they are clustered not with respect to their content or the specific ICT solution but with respect to the strategic vision embedded into the commitment.

Twenty-six projects address the community topic, regarding the social relationships of people living in city. ICT is especially used in its communication component, thanks to innovative technologies such as 3D and 4D interfaces, IoT, augmented reality and so on. These projects aim at improving the social capital of a smart city, thanks to the use of ICT to link people with each other and to enhance communication, information sharing and participation to social – even if virtual – events and discussions. These projects especially realise the smart city defined as virtual city or smart community.

Sixteen projects try to improve the quality of life, thanks to an improvement of the place where citizens inhabit, both supporting a shared way to live preserving the natural resources and enhancing projects to restore or improve the quality of buildings and other urban artefacts depicting the city environment.

Civil engagement regards the political participation of citizens to the city governance. In his projects, the role of the municipality is crucial, and it aims at improving the quality of life especially removing or reducing the negative aspects of the relationship between citizens and public administration. These projects try to increase trust in public bodies, to reduce bureaucracy and to improve the transparency and the accountability of public officials and politicians.

Another interesting impact of smart commitments regards the work-life balance. Today jobs especially in a city absorb a lot of time, like moving from home to the workplace. Smart projects can affect this aspect, both directly – for example, supporting e-working platforms – and indirectly, for example, reducing traffic, improving the efficiency of public transport and so on. No commitments explicitly declare to pursue this goal, but it is considered of primary importance in the OECD Better Life framework, and it emerges implicitly like a way to improve the quality of life, thanks to ICT in smart city programs.

Some projects are used to improve the quality of life in a city through the increasing of the urban security. These projects use technologies such as georeferencing, sensors and IoT. Some of them implement the concept of resilient city supporting risk prevention and risk management and also spreading the culture of resilience to actively involve citizens in preventing dangerous situations. The use of ICT and the active involvement of people are distinctive characteristics of smart projects deeply affecting the way to implement risk management in large urban areas. These projects have also a cultural role in changing the role of citizens in case of natural disaster or other dangerous events.

Other aspects of the Better Life Index are of less interest in smart projects. This survey contributes to trace a perimeter of the role of a smart city in really improving the quality of life of their citizens.

3.3 Discussion

The in-depth analysis presented above of the 42 ICT-based smart commitments focused on citizens is useful for two reasons.

The first reason is that the analysis provided an updated and systematic picture about the way European cities are implementing smart strategies to improve their citizens' quality of life. It emerges that, even if the European Union is strongly stimulating and supporting the smart city approach, its implementation differs significantly from country to country. Not only did different countries launch very different quantities of projects, but also the content of the projects is highly heterogeneous. This confirms the theoretical view of the smart city as a comprehensive strategy regarding all the aspects of the urban life and not just some topics (Chourabi et al. 2012). In addition, the theoretical view of the smart city as strongly based on citizens' participation emerges from a significant subset of projects, about 20% of the total number of implemented projects (Couclelis 2004).

The second reason is that the analysis presented in the previous sections shed light on how the smart city strategy, when implemented in specific projects, actually seeks to improve the city's quality of life. Indeed, both the theoretical definitions and the declared intents of municipalities committed in smart programs neglect to specify what a better life is and requires (Bowerman et al. 2000; Hodgkinson 2011).

Our investigation especially outlines two aspects.

The first regards the addressees of these subsets of smart projects. It emerges that 35 out of 42 projects – 85% – somehow address the generality of people living in a city. Thanks to the use of ICT, it is indeed possible to reach large amounts of people, involving actively the citizens not only in collecting data about their needs or expectations but also in the implementation processes of smart programs. Connecting people to each other, thanks to digital platforms and mobile apps flattens the city governance and generates a virtual but effective civil engagement that creates satisfaction and sense of belonging (Ricciardi et al. 2013).

The second aspect regards the way a smart city can affect the quality of life. The 42 smart projects collected in our subset are heterogeneous as for all the dimensions we investigate (smart city dimension, addressees, better life topics), but it emerges that ICT usage is a driver to collect different aspects in a comprehensive strategic vision about the city and the citizens' involvement. In fact, 33 out of 42 projects pursue the citizens' satisfaction not by addressing a specific aspect of their life but by addressing the multidimensional features of their life, including health, education, culture, environment, social relations and civil engagement. Benefits generated by these comprehensive smart commitments are not the sum but the product of the benefits produced by single applications (Dameri and Ricciardi 2015). The smart platform produces positive synergies that especially affect the intangible aspects of the quality of life in a city, creating a virtual space where communication, information sharing and participation are made possible, thanks to the joint effort of technological solutions and strategic planning (Nam and Pardo 2011; Larsen 1999; Lombardi et al. 2012).

This emerges not only from the quantitative analysis but also from the qualitative understanding of the storytelling introduced in the project data sheet. Each time it is possible to discover under the technological means the specificity of the urban fabric and the aim to positively transform the city life, thanks to a synergic combination of technology, people behaviour and participated territorial governance (Paskaleva 2009).

Almost all of the investigated projects are still in progress; therefore, it is not possible to give an assessment of their effectiveness in reaching the desired results. However, we firmly think that – grounding our opinion on both the theoretical definition of smart city and the storytelling emerging from all of the 42 projects – the merging of technology, governance and people is the strategic success factor of smart city commitments.

3.4 Focus on IoT

Even if our analysis is not especially focused on the different specific technologies implemented in each project, a special attention is devoted to IoT considered like a new technological driver to realise effective smart projects.

The quantitative analysis on the commitment database reveals that only 11 out of 366 commitments use IoT as IT technology to implement a smart solution; the number decreases to 3 out of these 11 if we concentrate our attention on the citizens' focus. It means that only 3 out of 42 projects focused on directly pursuing the citizens' well-being in a city are built on IoT as a technological infrastructure.

Examining these three commitments, we can better understand their specific aim and how IoT is used to implement a better life in urban areas.

- KnowHow is a project implemented in two UK cities that partnered (Southampton and Portsmouth) to instal large digital displays in the city centre. Thanks to a large number of different sensors, these information pods will be able to deliver real-time content to children, families, shoppers and tourists about the main street, with the aim to impact on their behaviours such as carrying goods, taking an alternative mode of travel and extending the shopping. The project is based also on open data and big data analytics and on gamification to engage the users in a full urban experience.
- Energy goals are a project implemented in Barcelona, Spain, aiming at achieving reduction in the energy consumption of public residential buildings through motivating learning and citizens' behaviour change. IoT is embedded in personalised tools using sensors, apps and social networks, delivered to the tenants; they are involved in engagement on energy savings, thanks to the real-time advices, alarms and information, learning algorithms and data consumption.
- Stakeholder platforms is implemented in Navarra (a Spanish region); it is both a physical and web-based platform creating an online open database of industries, clusters, SMEs, professionals and research centres involved in sustainable urban mobility. The platform aims at collecting and spreading relevant information through IoT and at promoting concerted actions in order to bring together integrated, scalable smart city solutions.

As it is possible to deduce from the analytical analysis of these three commitments focused on citizens and based on IoT, they share their orientation towards the data collection through sensors aiming at involving the citizens in changing their behaviour. The final goal is a more sustainable city, through sustainable mobility or energy consumption, but involving the proactivity of people instead of simply using data to connect things. This double layer of data communication – amongst devices and from devices to information platforms oriented to the citizens – is conceived like a way of capital importance to increase the positive effects of these projects on the city smartness.

4 Conclusions

The smart city idea is evolving towards an umbrella concept that implies the key role of ICT in enabling the city system's sustainability and resilience whilst pursuing citizens' quality of life.

Many cities are experimenting many different and idiosyncratic strategies to achieve these intertwining goals. This study provides the reader with a systematic map of these sometimes converging, sometimes diverging strategies. The results confirm that ICTs can actually serve as the unifying element that permits the integration and cross-fertilisation of many different initiatives in many different areas of urban life.

Currently, the main factor of possible smart city failure seems the poor integration between and across smart initiatives; therefore, ICTs and information systems management are in the position to play a crucial role for smart city success in the next future.

Possible future research steps include the dynamic monitoring of the impacts of smart initiatives, especially the intangible impacts relating to quality of life and the organisational challenges stemming from the need to leverage ICT to pursue the integration of smart project portfolios in different, specific contexts.

References

Anthopoulos, L., Fitsilis, P.: From digital to ubiquitous cities: defining a common architecture for urban development. In: Intelligent Environments (IE), 2010 Sixth International Conference on, pp. 301–306. IEEE (2010)

Bowerman, B., Braverman, J., Taylor, J., Todosow, H., Von Wimmersperg, U.: The vision of a smart city. In: 2nd International Life Extension Technology Workshop, Paris (2000)

California Institute.: http://smartcommunities.org/concept.php (2001)

Caragliu, A., de Bo, C., Nijkamp, P.: Smart cities in Europe. J. Urban Technol. **18**(2), 65–82 (2011)

Chourabi, H., Nam, T., Walker, S., Gil-Garcia, J. R., Mellouli, S., Nahon, K., Scholl, H. J: Understanding smart cities: an integrative framework. In: System Science (HICSS), 2012 45th Hawaii International Conference on, pp. 2289–2297. IEEE (2012)

Cocchia, A.: Smart and digital city: a systematic literature review. In Dameri, R.P., Rosenthal-Sabroux, C. (eds.) Smart City. How to Create Public and Economic Value with High Technology in Urban Space, pp. 13–43. Springer (2014)

Couclelis, H.: The construction of the digital city. Environ. Plan. B. **31**(1), 5–20 (2004)

Dameri, R. P.: Defining an evaluation framework for digital cities implementation. In: Information Society (i-Society), 2012 International Conference on, pp. 466–470. IEEE (2012)

Dameri, R.P.: Searching for smart city definition: a comprehensive proposal. Int. J Comput. Technol. **11**(5), 2544–2551 (2013)

Dameri R.P.: Using ICT in smart city. In: Smart City Implementation. Creating Economic and Public Value in Innovative Urban Systems. Springer International, Switzerland (2016a)

Dameri R.P.: Smart city definition, goals, and performance. In: Smart City Implementation. Creating Economic and Public Value in Innovative Urban Systems. Springer International, Switzerland (2016b)

Dameri, R. P., Cocchia, A: Smart city and digital city: twenty years of terminology evolution. In: ItAIS 2013, X Conference of the Italian Chapter of AIS (2013)
Dameri, R.P., Ricciardi, F.: Smart city intellectual capital: an emerging view of territorial systems innovation management. J. Intellect. Cap. **16**(4), 860–887 (2015)
Dameri, R.P., Rosenthal-Sabroux, C.: Smart city and value creation. In: Dameri, R.P., Rosenthal-Sabroux, C. (eds.) Smart City. How to Create Public and Economic Value with High Technology in Urban Space. Springer, Germany (2014)
Ergazakis, K., Metaxiotis, K., Psarras, J.: Towards knowledge cities: conceptual analysis and success stories. J. Knowl. Manag. **8**(5), 5–15 (2004)
European Parliament. Mapping Smart City in the EU. Brussels (2014)
Giffinger, R., Fertner, C., Kramar, H., Kalasek, R., Pichler-Milanović, N., Meijers, E.: Smart Cities: Ranking of European Medium-Sized Cities. Centre of Regional Science (SRF), Vienna University of Technology (2007)
Hodgkinson, S.: Is Your City Smart Enough? Digitally enabled cities and societies will enhance economic, social, and environmental sustainability in the urban century (2011)
Hollands, R.G.: Will the real smart city please stand up? Intelligent, progressive or entrepreneurial? City. **12**(3), 303–320 (2008)
Ishida, T.: Digital city of Kyoto. Mag. Commun. ACM. **45**(7), 76–81 (2002)
Komninos, N.: The architecture of intelligent cities: integrating human, collective and artificial intelligence to enhance knowledge and innovation. In: Intelligent Environments, 2006. IE 06. 2nd IET International Conference on, vol. 1, pp 13–20. IET. (2006)
Larsen, K.: Learning cities: the new recipe in regional development. OECD Observer No 217/218, Summer (1999)
Lombardi, P., Giordano, S., Farouh, H., Yousef, W.: Modelling the smart city performance. Innov. Eur. J. Soc. Sci. Res. **25**(2), 137–149 (2012)
Nam, T., Pardo T.A.: Smart city as urban innovation: Focusing on management, policy, and context. In: Proceedings of the 5th International Conference on Theory and Practice of Electronic Governance. ACM (2011)
Neirotti, P., De Marco, A., Cagliano, A.C., Mangano, G., Scorrano, F.: Current trends in Smart City initiatives: some stylised facts. Cities. **38**, 25–36 (2014)
OECD: OECD Better Life Initiative. OECD Publishing, Paris (2011)
Paskaleva, K.A.: Enabling the smart city: the progress of city e-governance in Europe. Int. J. Innov. Reg. Dev. **1**(4), 405–422 (2009)
Ricciardi, F., De Marco, M.: The Challenge of Service Oriented Performances for Chief Information Officers. In: Exploring Service science. Third International Conference, IESS 2012, pp. 258–270. Springer, Geneva (2012)
Ricciardi, F., Za, S.: Smart city research as an interdisciplinary crossroads: a challenge for management and organization studies. In: Mola, L., Pennarola, F. (eds.) From Information to Smart Society: Environment, Politics and Economics. Lecture Notes in Information Systems and Organisation, vol. 5, pp. 163–171. Springer (2014)
Ricciardi, F., Rossignoli, C., De Marco, M.: Participatory networks for place safety and livability: organisational success factors. Int. J. Netw. Virtual Org. **13**(1), 42–65 (2013)
Schuler, D.: Digital cities and digital citizens. In: Tanabe, M., van den Besselaar, P., Ishida, T. (eds.) Digital Cities II: Computational and Sociological Approaches, pp. 71–85. Springer, Berlin (2002)

Energy Consumption of the Building Sector: Incorporating Urbanization, Local Climate Change, and Energy Poverty

M. Santamouris and C. Cartalis

Abstract Addressing energy consumption in the building sector in Europe is considered a matter of urgency, taken its contribution to the emissions of air pollutants and greenhouses gases, heat release, and annual material and energy use. In this paper, it is shown that existing, business as usual scenarios for addressing energy consumption in the building sector underestimate such critical parameters as urbanization, local climate change, and energy poverty. Furthermore, it is shown that (a) the building stock cannot be separated from the space between and around the buildings, with the space being influenced and finally shaped by urbanization, and (b) energy poverty sets an upper limit with respect to the capacity of households to comply with local climate change and energy conservation objectives. Finally, the importance of the interlinks between energy consumption on the one hand and urbanization, local climate change, and energy poverty on the other is examined and demonstrated in view of proposing an integrated energy, environmental, and social policy for energy consumption in the building sector.

1 Introduction

Important sectoral policies have been defined and implemented in the developed world, aiming to improve the energy consumption of buildings, improve the environmental quality of cities and ameliorate the local climate change, and finally fight energy poverty and social disparities. Unfortunately, existing policies are usually fragmented and not combined with a wider and integrated framework where the economic, social, energy, and environmental issues are considered together. As a result, energy targets rarely consider the economic and social reality and stratification and usually define horizontal objectives neglecting the real needs and

M. Santamouris (✉)
School of Built Environment, University of New South Wales, Sydney, NSW, Australia

Physics Department, University of Athens, Athens, Greece
e-mail: m.santamouris@unsw.edu.au

C. Cartalis
Physics Department, National and Kapodistrian University of Athens, Athens, Greece

S.T. Rassia, P.M. Pardalos (eds.), *Smart City Networks*, Springer Optimization and Its Applications 125, DOI 10.1007/978-3-319-61313-0_8

the capacity of the weaker social and economic groups. It needs to be mentioned that energy objectives above the capacity and the technological state of the art of the local societies usually have low success, while they may reflect a solid risk of additional marginalization of segments of the population which are unable to follow demanding actions.

2 The Built Environment

The global characteristics of the built environment are determined by multilayered economic forces. Pressures related to the provision and accessibility of the basic needs; social aspects associated with local identity, culture, and security, in combination with economic burdens connected to personal and household income, labor market, and economic justice; and finally environmental stimuli linked to the use of land, materials flow, and energy production and delivery determine the characteristics of the built environment in a place (Smith et al. 1998).

The built environment is a complex notion; while it may be perceived as a simple collection of buildings in reality, it is the result of many multifaceted social and economic processes being mutually related to the prevailing microclimatic and environmental conditions. Energy and heat fluxes between buildings and the surrounding areas control the state of the thermal environment in the spaces between the buildings and thus affect their energy performance. Energy enters, passes, and leaves the urban system in several ways and in several physical states and forms. Fuels, electricity, radiation, and convective and latent heat are the main categories. To this end, the aspects of energy which are of interest depend on the particular scope of concern: urban planners, architects, city administrators, economists, statisticians, environmentalists, and physicists, each having a different focus (Chrysoulakis et al. 2013). For example, the interest of city administrators may relate primarily to the optimization of energy fluxes in order to address such questions such as how energy consumption can be controlled (e.g., guidelines for insulation of new houses, old-building retrofitting, etc.) or how an open space may develop lower air and surface temperatures. On the other hand, environmentalists are concerned with understanding how energy in its radiative, convective, and conducting forms is transferred and stored in the open environment, thus influencing the microclimate of the space between buildings. Often not all exchanges are addressed; for example, architects may concentrate on the building characteristics and underestimate the influence of the space between the buildings or the comfort felt by the inhabitants of the buildings. In any case, it is considered fundamental to accurately define the energy balance of the area concerned, by taking into consideration not only the building stock but also the open space. Rapid urbanization further complicates the situation, especially if it occurs without climate responsive guidelines. In this case, cities are progressively falling short of sustaining quality outdoor and indoor life. At the building level, such inadvertent climatic modifications have led to higher demand of urban energy resources. In order to address and reestablish energy consumption of the building sector in Europe, it is important that urban spaces are made comfortable as far as the ambient climate permits.

Finally, to analyze energy–material exchanges and provide parameter values, ground-based measurements have been excessively used. In some cases, however, the density (temporal and spatial) of ground-based measurements may be limited, a fact which limits the accuracy of analysis. To circumvent this limitation, Earth observation (EO) and geographic information systems (GIS) have been proven useful. EO and GIS provide data and analytical tools to assess sustainable urban planning and management (Nichol et al. 2007), radiative exchanges (Chrysoulakis, 2003; Rigo and Parlow, 2007; Polydoros and Cartalis, 2014), surface characteristics (Stathopoulou and Cartalis, 2007), surface turbulent sensible and latent heat fluxes (Xu, et al. 2008), climate change mitigation, and urban heat island (Cartalis et al. 2015; Dousset et al. 2011) and thus contribute to the assessment of urban metabolism components (Chrysoulakis et al. 2013). As planners need to consider environmental and socioeconomic issues and impacts simultaneously, evaluation methods and tools need to be integrated in a GIS environment to address multiple aspects within decision-making regarding sustainable urban planning.

3 Critical Dimensions: Urbanization, Local Climate Change, and Energy Poverty

3.1 Urbanization

Nearly 73% of the European population lives in cities, and this is projected to reach 82% in 2050. Urbanization is associated with changes in materials and energy flow in the cities (Chrysoulakis et al. 2013), can increase pressures on the environment, may alter the microclimate and may support climate change mechanisms through enhanced emissions of greenhouse gases. It may also reduce the resilience of cities to extreme events associated to climate change due to urban land take and the limited capacity to regenerate the building stock of the cities.

The interplay between urbanization and local climate change needs to be assessed carefully. According to several researchers (Santamouris et al. 2014), the long-term trend in surface air temperature in urban centers is associated with the intensity of urbanization. Urbanization is also linked to urban sprawl as emissions of greenhouse gases are higher in commuter towns not only because of car dependency but also due to the characteristics of the buildings. Typically, European cities are dense but they are becoming less dense at their boundaries, whereas large differences are observed in the level of sprawl both between and within European countries.

3.2 Local Climate Change

Local climate change defined as the urban heat island phenomenon (UHI) is a major threat to human well-being. It leads to higher urban temperatures compared to the surrounding rural or suburban areas because of the positive thermal balance of cities created by the production of anthropogenic heat, the high absorption of solar radiation by the urban fabric, and the reduced cooling intensity of the low-temperature atmospheric heat sinks (Santamouris 2001). The urban heat island phenomenon is extremely well documented with measurements of near-surface air temperature in more than 400 major cities in the world (Santamouris 2015a), with its average intensity varying between 3 and 6 K; it has been also documented through estimates of land surface temperatures from satellite measurements (Cartalis et al. 2015). The extent and intensity of the urban heat island depend on the presence of anthropogenic heat sources as well as on land use/land cover. The urban heat island increases energy consumption of buildings for cooling purposes, deteriorates the indoor and outdoor comfort conditions, raises the concentration of harmful pollutants like the tropospheric ozone, and has a serious impact on human health and mortality (Santamouris 2015b).

3.3 Energy Poverty

Energy poverty is defined as "the situation in which a household lacks a socially and materially necessitated level of energy services in the home" Bouzarovski (2014). According to the (European Commission: 2015), almost 11% of the European population in 2012 was not able to meet the basic heating requirements, while 19% of the population faced an overheating problem in summer. Energy poverty affects seriously the quality of life of lower-income households, while it has a very considerable impact on health and mortality rates (Kolokotsa and Santamouris 2015).

4 The Challenge Ahead

Urbanization, local climate change, and energy poverty will define the future priorities and will drive the evolution of the building sector. The adoption of a proactive agenda for the building sector is necessary; such agenda needs to be based (a) on the integration of economic, social, environmental, and technological criteria in areas where the major economic and social disparities are recognized and (b) the achievement of strong synergies, links, and positive trade-offs with the major technological and economic mega trends in the world, like the advanced ICT technologies, big data (including Earth observation data), and smart products and technologies.

The present paper investigates and discusses the possible limitations and inconsistencies of the current energy policies for the building sector in relation to the social, economic, microclimatic, and environmental problems of our cities. The overall aim of the paper is to demonstrate the need for an integrated policy addressing the following (research) questions:

(a) To which extent policies aiming to implement intensive energy conservation and near zero energy solutions in individual buildings can face and overcome the existing environmental, social, economic, and technological barriers and limitations and consist an efficient tool to improve the global environmental quality of the urban built environment?
(b) How can we link policies addressing the energy consumption of the building sector to the microclimatic and environmental conditions of the space between them?
(c) Which interventions should be planned and implemented to respond in a global and integrated way so as to improve the energy, environmental, social, and economic reality of our cities?

5 Overheated and Deprived Cities as Promoters of Social, Economic, Energy, and Environmental Problems

Global and local climate change increase the near-surface ambient temperature of cities and rise the frequency of heat waves (Kapsomenakis et al. 2013). Increased ambient temperatures have a very significant impact on the overall energy and environmental quality of cities, affecting indoor and outdoor comfort conditions, augmenting the concentration of harmful pollutants like tropospheric ozonetropospheric ozone, and finally increasing the peak electricity demand and the energy consumption of buildings during the summer period.

In Europe alone, the urban heat island phenomenon was recognized in 110 cities, on the basis of temperature measurements performed either using standard or nonstandard meteorological stations, mobile traverses, or satellite observations. Existing works report either the annual average, the average maximum, or the absolute maximum urban heat island intensity. The intensity of surface urban heat island ranges between 1 and 6 K, depending, among others, on the geographic location and land use/land cover characteristics.

The above analysis demonstrates the importance of the urban heat island in Europe in terms of the increase of ambient temperature. In particular, the elasticity of the electricity power demand regarding the ambient temperature has been studied for 17 European countries, Austria, Belgium, Denmark, Finland, France, Germany, Greece, Hungary, Ireland, Italy, Luxemburg, the Netherlands, Norway, Portugal, Spain, Sweden, and the United Kingdom (De Cian et al. 2007). It is found that in Southern European countries, the average elasticity is close to 1.54% while for the mild and cold countries is 0.54% and 0.51 %, respectively. The increase of the peak electricity load per degree of ambient temperature increase varies as a function of the characteristics of the local energy system and the quality of the buildings. An analysis provided by Santamouris et al. (2015), for several cities in the world, concluded that the increase of the peak electricity demand per degree of temperature increase is ranging between 0.45% and 4.6%. In parallel, it is reported that the increase of the total electricity demand per degree of temperature increase is close to 1.6% in Spain, 1.1–1.9% in Greece, and 0.5% in the Netherlands (Santamouris et al. 2015).

Several studies in Europe have attempted to identify the energy penalty induced by the urban heat island effect. Kolokotroni et al. (2007) found that because of the UHI, the cooling load of buildings in central London increased between 27% and 45%, compared to the adjacent rural areas. Fanchiotti and Zinzi (2012) reported that the increase of the cooling load of buildings in Rome induced by the UHI phenomenon was close to 130%, compared to the rural reference zones. A similar increase of the cooling load of buildings, 120%, is calculated by Santamouris et al. (2001) for downtown Athens, Hassid et al. (2000) found that the cooling demand of buildings in western Athens increases by 67% because of the urban heat island, and Polydoros and Cartalis (2014) showed the impact of urban sprawl to the thermal environment – and consequently to energy consumption – for Athens. Santamouris (2014a) found that the average increase of the cooling energy demand in the period 1970–2010 for many European cities was 23%, while the corresponding decrease of the heating requirements was 19%. Globally, it is found that local and global climate change increases the total energy consumption of the buildings by 11%.

Higher urban temperatures have a strong impact on the environmental quality of cities, increasing the concentration of pollutants, deteriorating indoor and outdoor comfort conditions, and increasing the global ecological footprint of cities. Studies in Athens, Greece, show that the ecological footprint of the city increased by 1.5–2 times, mainly because of the urban heat island (Santamouris et al. 2007). In parallel, studies performed in Athens and Paris, France, showed a very strong correlation between the increased ambient summer temperatures and high concentrations of tropospheric ozone exceeding the permitted thresholds (Sarrat et al. 2006; Stathopoulou et al. 2008).

Poor environmental and economic conditions in the urban environment may strengthen the intensity of these problems and promote further energy, economic, and environmental deprivation. According to European statistical data, almost 10–25% of the European population is living in low-income households (Dol and Haffner 2010). Low-income urban population is in many cases unable to cover its basic energy needs because of the serious lack of economic resources, the considerably high price of fuels, and the low energy and environmental quality of the houses they use to live (Santamouris and Kolokotsa 2015). According to the official European data, almost 73% of the low-income population in Portugal and about 15–25% of the low-income population in the Southern European countries can not afford to pay for heating (European Foundation for the Improvement of Living and Working Conditions 2003). This results in unacceptable indoor environmental conditions that adversely affect health and quality of the life of lower-income households. In fact, low-income population is exposed to extreme indoor temperature conditions both during the summer and winter period. Measurement of the indoor environmental quality in low-income houses in Athens during the winter period showed that the average indoor temperature was several degrees below the comfort zone, while during periods of significant cold weather, indoor temperature was below 10 °C (Santamouris et al. 2014). Numerous studies performed in the United Kingdom have also found that winter indoor temperatures in low-income houses are substantially below the thermal comfort thresholds (Yohanis and Mondol 2010; Hutchinson et al. 2006).

Extreme indoor and outdoor temperatures have a serious impact on the health of low-income urban population. Studies have concluded that urban population living in low-income urban zones with low economic ability presents a much higher health risk during extreme weather conditions (Keatinge et al. 2000; Basu and Samet 2002). In parallel, social research has concluded that vulnerability of low-income households is aggravated in poorer economic urban zones presenting a limited capacity for improvements (Mc Gregor et al. 2007), while mortality rates are higher in urban areas presenting a high deprivation score (Middelkoop 2001).

6 Energy Consumption in the Building Sector in Europe: Interrelations and Synergies with Environmental and Social Issues

6.1 The Building Sector

The building sector consists the biggest energy consumer in Europe and is responsible for about 41% of the total energy consumption. Space heating is the most energy consuming use, 71%, followed by air conditioning and other appliances 15%, water heating 12%, and cooking 4% (Odyssee-Mure 2012). Energy consumption is constantly increasing in Europe; in particular, energy consumption of the residential sector has increased by 14% between 1990 and 2012, while the tertiary sector has a continuous increase that varied between 0.6% for the period 1990–2000, 1.2% for the next 10 years, and 1.1% for the period 2010–2020 (EU-25 Energy and Transport Outlook to 2030 2015). Increase of the energy consumption of the residential sector is mainly attributed to the important increase of the households in Europe, while the corresponding increase of the tertiary sector is attributed to the significant evolution of the services sector.

Intensive energy reduction policies are defined and implemented by the European Union mainly during the last 20 years. The European Directive on the Energy Performance of Buildings has set specific targets and objectives for all member states, including the obligation that all new buildings present an almost zero energy consumption by 2018 for the public and 2020 for the rest of the buildings. Additionally, the European Commission has implemented the Directive on the

Energy Efficiency that requires that all member states have to implement mandatory energy measures to achieve energy savings of 1.5% per year. Finally, the European Commission has defined and implemented the Renewable Energy Directive, the Ecodesign Directive, and the Energy Labelling Directive that contribute toward a reduced consumption of conventional energy fuels in the built environment.

As a result of the technological and legislative developments, the energy consumption of the new buildings in Europe presents almost 40–60% lower energy consumption compared to buildings constructed before 1990. Given the relatively small number of new buildings in Europe and the focus of the efforts on heating systems, the energy consumption spent in Europe for heating purposes is decreasing at a rate close to 1.4% per year and dwelling (Odyssee-Mure 2012). On the contrary, the cooling energy consumption is increasing at a very high rate especially in the European South, mostly due to local climate change as reflected by the presence of the urban heat island. It is characteristic that in specific countries like Bulgaria, the energy consumption for cooling has increased by 100%, while the corresponding increase in Spain and Italy is 30% (Odyssee-Mure 2012).

It is important to note that economic turmoil and crises affect mainly the lower- and median-income population and decrease the energy consumption because of the income decrease. The financial crisis of 2007–2012 has resulted in a 4% decrease of the household energy consumption in Europe, while in selected countries heavily touched by the crisis, like Portugal and Ireland, the corresponding decrease was around to 16% and 22%, respectively (Odyssee-Mure 2012).

6.2 Environmental Issues: The Role of Local and Global Climate Change

Local and global climate change have a serious impact on the actual and the projected energy consumption of urban buildings. As already mentioned above, local climate change increases significantly the energy demand for cooling and decreases slightly the energy demand for heating purposes. The combination of local and global climate change increases the energy stress in cities and individual buildings much more than the background impact of each of the two phenomena (Li and Bou Zeid 2013). Simulations of the future energy consumption of buildings in Europe considering various climatic scenarios concluded that by 2050 the cooling energy demand may increase between 40% and 70%, while the expected decrease of the heating demand may be limited between 9% and 17%.

Buildings may generate 2300–5000 Wh/m^2 of anthropogenic heat per day (Hamilton et al. 2009) and may increase the magnitude of the urban heat island by 1.5 K (Bohnenstengel et al. 2014). The extensive use of renewable energies in buildings to satisfy the requirements on zero energy consumption may contribute highly to the reduction of the emission of greenhouse gases. Studies on the mitigation potential of building-integrated photovoltaics in Europe estimate that may contribute to a decrease of the greenhouse emissions as much as 10% of the total emissions by 2010 (Defaix et al. 2012).

Although energy conservation policies contribute significantly to ameliorate local and global climate change, the actual energy penalty induced to the building sector because of the actual temperature increase in cities is very significant. As already mentioned, increased ambient temperatures affect both the required electrical power and the absolute energy demand. According to Santamouris (2014a), an assessment of the existing impact of urban heat island on the energy consumption of buildings shows that the global energy penalty per square meter of city space and per degree of the UHI intensity is close to 0.74 ($\pm$0.67), while the global energy penalty per person is found to be in 237 ($\pm$130) kW h/p. Increased summer temperatures will affect the sizing of the HVAC systems installed in buildings and will decrease their coefficient of performance. Several studies have shown that the necessary cooling capacity in urban zones suffering from a serious temperature increase may increase by 100 %, while the COP of the conventional HVAC systems may decrease up to 25% (Hassid et al. 2000). This will result in an increased capital and running cost of the HVAC systems and will burden additionally the urban population.

Increased ambient and water temperatures may affect significantly the capacity of the power plants in Europe. Studies considering various climatic scenarios estimate that the capacity of the power plants may decrease between 6.3% and 19% during the summer period (VanVliet et al. 2012), and the expected missing power capacity may vary between 6 and 19 GW (Rubbelke and Vogele 2011). Given the necessary new investments to ensure the high reliability of the power network in Europe, prices of electricity may increase up to 80% in selected countries (Rubbelke and Vogele 2011). Furthermore, climate change has a serious impact on the cooling potential of natural sinks decreasing significantly the potential of passive cooling techniques. Recent analysis of the European climatic data shows that the number of warm nights in various parts of Europe has increased considerably (EURO4M 2015), while the cooling potential of night ventilation techniques has been seriously affected (Santamouris and Kolokotsa 2013). Experimental studies have revealed that in air-conditioned buildings, the cooling capacity of the night ventilation techniques has been reduced up to 35% (Geros et al. 1999).

6.3 Social Issues

The relationship between social parameters and the energy consumption in the built environment is quite complex. It is evident that energy conservation measures and buildings of higher energy quality improve the quality of life of low-income households and contribute to improve their economic conditions. Households of lower income cannot satisfy the basic energy needs as usually they live in dwellings of poor environmental quality presenting a very high energy consumption (Kolokotsa and Santamouris 2015). Programs involving energy-efficient rehabilitation of low-income houses contribute highly to escape from the energy poverty. Evaluation of the rehabilitation program of low-income housing in the United Kingdom concluded that a high fraction of the beneficiaries has escaped from energy poverty (Preston et al. 2008).

In parallel, implementation of energy-intensive measures in low-income households presents important non-energy benefits. In particular, it creates new economic opportunities, offers new labor opportunities, increases the national Gross Domestic Product (GDP), decreases the emission of greenhouse gases, increases the economic value of dwellings, increases the social status of the deprived population, and, most importantly, improves indoor comfort conditions and protects the health of the low-income population. Buildings with proper ventilation rates present superior indoor air quality levels that result in about 50% reduction in health problems (Carnegie Mellon 2005) and a significant decrease of hospital visits. It is characteristic that in the United Kingdom, the energy benefit of the national health system for each pound invested in rehabilitation of low-income housing is close to 0.42 lb (Chief Medical Officer 2009).

However, specific attention has to be given so as to avoid that the promoted energy performance targets do not exceed the state of the art, in particular, in the less-developed and possibly deprived zones of Europe. In such a case, the concerned population will not have access to the required credit and capital investments, and the final energy result will not satisfy the targets and requirement and may enhance further and economic discrepancies in Europe resulting in the energy-poor population to become highly marginalized in economic and social terms. According to several studies, this is very likely to happen in Europe if the necessary precautions are not taken on board and if maximalistic energy targets are considered. In particular, it is estimated that while the potential energy conservation in Europe will reach 72% by 2050 as compared to the 2005 levels, the possible unrealized energy benefits may reach 46% (Urge-Vorsatz et al. 2012).

7 Toward an Integrated Energy, Environmental, and Social Policy for the Building Sector

Given the strong interrelation of economic, environmental, and social issues for defining the energy performance of the built environment, an integrated approach for energy conservation for the building sector in Europe is strongly required. Such an approach should have the following objectives:

(a) Decrease the energy consumption and improve the environmental quality of the building sector, mitigate the local and global climate change, and fight energy poverty

(b) Promote measures that increase the added value in the local societies, generate wealth for the population, create new opportunities in the labor market, and decrease the energy and environmental burden of the population
(c) Protect the population from intrinsic and extrinsic risk factors and decrease the exposure and hazard levels in the society
(d) Distribute public and market wealth in a fair just way manner between the population in order to decrease socioeconomic inequalities that result in energy and environmental poverty

7.1 New Materials, Systems, and Technologies

Use of state-of-the-art technologies may reduce the energy consumption of new buildings for heating and cooling up to 90% compared to a conventional building. New state-of-the-art energy-efficient buildings designed using an advanced envelope of energy management, ventilation, and heating technologies may present a final heating energy consumption around to 15 kWh/m^2/year irrespective of the local climatic conditions. Such a consumption is four to five times lower than that of a new building and about 15–25 times lower than the consumption of an existing building (Harvey 2010). Passive cooling technologies involving the use of reflective coatings or the building envelope, solar cooling systems, natural ventilation, and use of environmental heat sinks for heat dissipation may reduce the cooling needs up to a minimum. In particular, the use of cool roofs in poor energy performing buildings may result in very significant energy reductions even in the north of Europe (Synnefa et al. 2007; Mastrapostoli et al. 2014). Implementation of cutting-edge ICT and communication technologies may also contribute to decrease the electricity consumption of the building equipment by 65% by 2020 compared to the existing average situation (Urge-Vorsatz et al. 2012). In particular, the use of smart meters as well as of the smart grids can optimize the use of energy in buildings, facilitate the integration of renewable energies, and minimize the possible power interruption. As reported, power interruptions in US buildings cost about 80–100 billion US dollars per year (Hamachi LaCommare and Eto 2006).

State-of-the-art energy-efficient buildings present almost 28% lower environmental impact than a conventional design (Thiers and Peuportier 2008). Although the common belief is that a state-of-the-art new energy-efficient building presents a much higher investment cost, the actual data shows that the additional cost compared to a conventional design may be around 5–8%. Other studies estimate that the possible additional investment cost of new buildings is almost negligible or even negative given that the additional investments on the building's envelope are counterbalanced by the substantially lower size and cost of the energy supply system (McDonell 2003). However, this is not the case for existing buildings that have to be retrofitted to improve their energy performance. The specific retrofitting cost may depend on the characteristics of the building and the local economic conditions; however, as mentioned in Zerohomes (2015), it may not be lower than 50,000 US$

per house. Reports from the Netherlands on the retrofitting of social housing into almost near zero energy buildings concluded that the retrofitting cost is close to 88,000 US$ per house, but it is decreasing continuously and may be as low as US$ 45,000 in the near future (Coexist 2015). A very recent study of the European Joint Research Center (JRC 2015) concludes that energy retrofitting technological solutions may be considered as economically feasible if the cost is below 300 euros per square meter in European countries with a well-established property market and below 500 euros in those countries with less mature markets. The same report points out that energy-efficient renovation may not be affordable for European citizens in countries presenting a GDP below the European average (JRC 2015). Therefore, it is a real challenge for the European citizens and the states to determine how "deeply" they have to retrofit buildings and also how the necessary cost will be covered.

7.2 Smart Cities

Cities in the future need to enhance aspects related to the urban form and density, reduce significantly the energy for transport, implement energy-efficient urban energy systems for the building sector, optimize activity patterns using advanced ICT technologies, and minimize the energy consumption of buildings and the related greenhouse gas emissions through the use of advanced energy technologies. A city implementing the above may be clearly defined as a smart city. However, smart low-carbon cities remain a noble and inspirational technological objective. The cost associated with the development of smart, low- or zero-carbon cities like the Masdar City in UAE is prohibitively high for the European countries. As mentioned by GEA (2012), future cities of the world should integrate by 2050 almost three billion additional citizens, and the cost of a zero-carbon scenario could cost US$ 1000 trillion that are equivalent to the world GDP for 20 years.

7.3 Countermeasures to Local and Global Climate Change

Important countermeasures to fight local and global climate change have been developed and tested in many real scale projects during the last years. Among the most developed ones are those aiming to increase the albedo of the cities, augment the urban green in cities and buildings, decrease the anthropogenic heat, control solar radiation through efficient shading of outdoor spaces, and use of environmental sinks of lower temperature like the water and the ground (Santamouris and Kolokotsa 2016).

In terms of mitigation technologies, the use of advanced materials for buildings and open spaces has gained the highest acceptance and presents the most commercial mitigation technology. Use of reflective materials in the urban fabric can contribute highly to increase the albedo of cities. Materials presenting a high reflectivity together with a high emissivity value may be used on roofs and pavements of the cities (Kolokotsa et al. 2013). Several technologies of reflective, cool materials have been developed and are available in the market, including ultrawhite, infrared-reflective, nanotechnological doped, retroreflective, and thermochromic coatings (Santamouris et al. 2011). Most of these materials have been incorporated into composite products like pavements, asphalt, and roof coatings and used in thousands of existing applications (Synnefa and Santamouris 2012). Especially, several cool pavement technologies have reached a very advanced technological stage and present a very credible mitigation solution able to decrease ambient temperature and improve comfort. Extensive experimental testing of these composite products has shown that they present a much lower surface temperature, up to 15 K, than conventional materials of the same color (Synnefa et al. 2011). Cool pavements and roofing materials have been used in hundreds of real large-scale projects aiming to decrease the outdoor ambient temperature. An evaluation of the existing monitoring results collected from almost all projects where monitoring data are available concluded that when the city's albedo is increasing by 0.1, the average summer ambient temperature may decrease by 0.3 K, while the average decrease of the peak summer ambient temperature is close to 0.9 K (Santamouris 2014b). The use of reflective materials in cities has a substantial energy benefit for the individual buildings and the whole built environment. Most of the existing applications concluded that cool roofs applied in individual buildings may reduce its cooling demand by 20–40 % (Synnefa et al. 2012b), while the increase of the city albedo may decrease considerably the energy demand of the building stock in the concerned urban zones (Schiano-Phan et al. 2015). Cool pavement technologies applied in the urban environment help to improve outdoor thermal comfort conditions and increase the visiting capacity of the spaces (Gaitani et al. 2011). In parallel, they may considerably contribute to the protection of local populations and decrease the exposure and the risk of the vulnerable population during extreme heat events like heat waves (Santamouris et al. 2008) while helping to reduce atmospheric pollution and, in particular, the exposure to tropospheric ozone (Stathopoulou et al. 2008).

Mitigation technologies based on the use of reflective materials have a relatively low cost and can be an excellent alternative for the climatic rehabilitation of deprived urban areas. The additional cost of cool roofs is between 0.5 and 2 US$ per square meter (Cool Roofs Rating Council 2015), while analysis of the results of many European projects involving cool pavements showed that the additional cost varies between half to one Euro per square meter (Fintikakis et al. 2011; Santamouris et al. 2012b).

7.4 Fighting Energy Poverty

Fighting energy poverty and protecting the vulnerable population are a major obligation of our societies. Low-income houses are characterized by low-energy performance and not acceptable indoor environmental quality, while the households can't afford to cover their energy needs. To face the problems, important policy and technological tools are designed and implemented in the various European countries, such as the retrofitting of the low-income housing stock as well as the retrofitting of open spaces of deprived zones in order to improve their environmental quality and upgrade the social use.

Experience from an intensive energy-efficient rehabilitation of low-income houses in the United Kingdom showed that 70% of the recipient households were able to move from energy poverty conditions (Preston et al. 2008). Selection and implementation of energy-efficient measures in low-income houses are subject to budget restrictions. Most of the existing interventions aim to upgrade the energy class of the houses up to the minimum acceptable level and are based on the use of well-known technologies like envelope insulation, efficient windows, and high-performance heating systems. The achieved reduction of the energy consumption in the low-income, rehabilitation program of the United States ranges between 25% and 40% per dwelling, and the corresponding average cost was close to 2700 US$ per dwelling (Enterprise Community Partners, Inc 2008). Much higher rehabilitation budgets, close to 50,000 US$/dwelling, have to be spent in order to achieve energy reductions, higher than 50% (Enterprise Community Partners, Inc 2008). However, data from the SOLANOVA social housing retrofitting project carried out in Hungary showed that the energy consumption of these buildings can be reduced up to 80% at a cost close to 250 euros per square meter + VAT (Hermelink 2006).

Protection of the health of the residents of low-income households is the major benefit arising from the retrofitting of low-income dwellings. It is characteristic that only in the United Kingdom the estimated additional health cost of cold homes is close to 1.36 lb per year (Marmot Review 2011), while as already mentioned almost 40% of the budget spent on the retrofitting of the low-income houses returns back to the national economy as an economic benefit from reduced health care (Chief Medical Officer 2009). In parallel, the expected environmental benefits may be very significant. Estimations showed that only in the United Kingdom the expected CO_2 reductions from the retrofitting of the low-income housing are close to 24 $MTonCO_2$ that is equivalent to 33% of the emissions of the national transport sector (Washan et al. 2014). Finally, important financial benefits result from the energy upgrade of low-income houses. As mentioned by Washan et al. (2014), there is an increased GDP benefit of 3.2 lb per pound invested in the energy efficiency of low-income homes and 1.27 lb arising from additional taxes.

8 The Economic Dimension of the Building Sector

Economic activities in Europe around the energy retrofitting of buildings were close to 283 billion Euros by 2011, out of which 166 billion were related to envelope technologies (Joint Research Center 2015). It is estimated that the implementation of state-of-the-art energy technologies for all buildings in the world requires investments of about US$ 14 trillion, by 2050, while the corresponding energy savings are estimated close to US$ 58 trillion (GEA 2012). The corresponding investment cost for Western and Eastern Europe is close to 2.5 and US$ 0.35 trillion, respectively, while the expected energy benefit is estimated to be around US$ 11 trillion for Western Europe and US$ 2.5 trillion for Eastern Europe (GEA 2012). The investment cost and the energy benefits of several building energy retrofitting scenarios are examined by the Building Performance Institute of Europe (2011), for the whole European building stock. It is estimated that the required annual investments by 2050, in present value, to achieve an almost zero energy renovation level are close to 937 billion euros, while the expected energy gains are close to 1.3 billion euros.

A recent report by Pike Research (2012) estimates that the worldwide revenue from zero energy buildings will reach almost US$ 690 billion by 2020 and will increase up to 1.3 trillion by 2050, presenting an annual growth rate close to 4.3%. According to the report, a very high fraction of the growth will occur in Europe. Another study of a similar nature (Navigant Research 2015) estimated that the annual world revenue from zero energy buildings will reach US$ 1.4 trillion by 2035, while the total revenue is expected to be close to US$ 4.6 trillion by 2035. Finally, the International Energy Agency (2014) estimates that the expected investments in energy efficiency of the building sector may reach US$ 2.33 trillion during the period 2014–2035. Cutting-edge building technologies present the highest potential for future market development and financial benefits. The market value of smart metering technologies increased up to US$ 10 billion per year in 2012 compared to 2.5 billion in 2009 and may reach US$ 20 billion by 2020, while the smart grid market may worth up to US$ 200 billion by 2020 compared to US$ 100 billion by 2015. In parallel, the world market in glass products for buildings was close to US$ 83 billion by 2014 (Visiongain 2014), and the market of phase change materials used for latent heat storage may reach US$ 1.76 billion by 2020.

9 The Social Dimension of the Building Sector

The building sector in Europe employs almost 11.5 million of people, while specialized construction that includes energy retrofitting provides almost 7.84 million of jobs (Joint Research Center 2015). Implementation of advanced energy efficiency technologies in the building sector may generate a very high number of new additional jobs. Recent studies evaluating the labor potential of energy efficiency measures in the building sector of nine European countries concluded that there are between 4.0 and 12.3 jobs per million of US$ invested (Joint Research Center 2015). Considering that GEA (2012) estimates that the necessary annual investments to achieve a low-carbon building stock in Europe by 2050 are close to 937 billion euros, it may be easily estimated that the expected saved or created employment should be between 3.7 and 11.5 million of additional jobs.

There are not available estimates about the required investments to eradicate energy poverty in Europe. In the United Kingdom, it is estimated that the necessary investments to improve the energy efficiency of the 2.5 million homes of the energy-poor households, up to the threshold level to escape energy poverty, are close to 4.6 billion pounds or 1840 lb per house (Preston et al. 2008). Another study estimated that the required capital investments to decrease by 50% the energy consumption of 550,000 low-income houses in the United Kingdom are between 3.9 and 17.5 billion pounds, which correspond to a cost per house between 7100 and 31,800 lb (Jenkings 2010). Based on the above figures, and considering the construction costs and the number of energy poor in the various European countries, the estimated cost to improve the energy efficiency of the 150 million energy poor in the European Union countries (Bird et al. 2010), up to the limited threshold to escape energy poverty, should be between 130 and 150 billion US$. In parallel, the cost of a possible deep energy efficiency rehabilitation to achieve a 50% energy reduction may range between 400 billion US$ and 2.0 trillion US$. Investments aiming to increase the energy efficiency of low-income households may create significant employment opportunities for the lower-income communities. The US Department of Energy estimated that for every million of US$ invested in the energy refurbishment of low-income houses, almost 52 additional jobs are generated for the local low-income population (USDOE 2006).

10 Conclusions

Energy consumption in the building sector in Europe should not be only considered as being dependent on the quality of the building stock. It also depends, although at varying degrees, on such issues as urbanization, local climate change (especially in light of the expanding presence of the urban heat island phenomenon in European cities), and energy poverty. Failure to take into consideration any of the issues above will inevitably result in incomplete policies and may even increase energy consumption and social disparities. In this paper, the need to take into consideration the microclimatic, environmental, and social characteristics of the space around and between the buildings has been shown to be equally critical as the type of the building stock itself. Furthermore the need for an integrated energy, environmental, and social policy for the building sector has been demonstrated. The main objectives of this policy are to improve the microclimate in the space around and between the buildings, to improve the environmental quality of the building sector, to mitigate the local and global climate change and to fight energy poverty, and to finally lead to sustainable energy conservation. The policy has been shown to contribute to the creation of new jobs, to deliver significant economic gains for the market, and to provide investment opportunities in Europe, a fact which of paramount importance taken the financial stagnation observed in several member states. Finally, and most importantly, such a policy carries the potential to ameliorate social disparities which stem from energy poverty as well as due to the fact that poor building stock is usually associated with low-income citizens. It is for the benefit of the European Union to address the issue of energy consumption in the building sector in such a multifaceted approach, instead of concentrating its effort in measures which are exhausted in improving the technical standards of the building stock.

References

Akbari, H., Cartalis, C., Kolokotsa, D., Muscio, A., Pisello, A.L., Rossi, F., Santamouris, M., Synnefa, A., Wong, N.H., Zinzi, M.: Local climate change and urban heat island mitigation techniques – the state of the art. J. Civ. Eng. Manag. **22**, 1–16 (2015.) In Press

Basu, R., Samet, J.M.: Relation between elevated ambient temperature and mortality: a review of the epidemiologic evidence. Epidemiol. Rev. **24**, 190–202 (2002)

Bird, J., Campbell, R., Lawton, K.: The Long Cold Winter: Beating Fuel Poverty. Institute for Public Policy Research and National Energy Action, London (2010) www.vhscotland.org.uk/library/misc/The_Long_Cold_Winter.pdf

Bohnenstengel, S.I., Hamilton, I., Davies, M., Belcher, S.E.: Impact of anthropogenic heat emissions on London's temperatures. Q. J. R. Meteorol. Soc. **140**(679), 687–698 (2014)

Bouzarovski, S.: Energy poverty in the European Union: landscapes of vulnerability. WIREs Energy Environ. **3**, 276–289 (2014). doi:10.1002/wene.89

Building Performance Institute of Europe: Europe's Buildings under the Microscope. http://www.europeanclimate.org/documents/LR_%20CbC_study.pdf (2011)

BWI, Building and Wood Workers International: Briefing on Labour Standards in Construction Contracts presented at the BWI World Council 7th December 2006 (2006)

Carnegie Mellon: University Center for Building Performance. As cited in Greening America's Schools: Costs and Benefits. October 2006. G. Kats, Capital E. Available: http://www.usgbc.org/ShowFile.aspx?DocumentID=2908(2005). Accessed 24 May 2010

Cartalis, C.: Towards resilient cities – a review of definitions, challenges and prospects. A. Build. Energy Res. **8**, 259–266 (2014)

Cartalis, C., Polydoros, A., Mavrakou, T.H., Assimakopoulos, D.N.: Use of earth observation for the development of resilience and adaptability plans for the thermal environment in urban areas, manuscript submitted for publication. Open. J. Remote Sens. (2015)

Chief Medical Officer: Annual Report. http://www.sthc.co.uk/Documents/CMO_Report_2009.pdf (2009)

Chrysoulakis, N.: Estimation of the all-wave net radiation balance in urban environment with the combined use of Terra/ASTER multispectral imagery and in-situ spatial data. J. Geophys. Res. **108**(D18), 4582 (2003)

Chrysoulakis, N., Lopes, M., San José, R., Grimmond, C.S.B., Jones, M.B., Magliulo, V., Klostermann, J.E.M., Synnefa, A., Mitraka, Z., Castro, E., González, A., Vogt, R., Vesala, T., Spano, D., Pigeon, G., Freer-Smith, P., Staszewski, T., Hodges, N., Mills, G., Cartalis, C.: Sustainable urban metabolism as a link between bio-physical sciences and urban planning: the BRIDGE project. Landsc. Urban Plan. **112**, 100–117 (2013)

Coexist: In just a week, this kit turns old houses into zero-energy homes (for free). Available through: http://www.fastcoexist.com/3046525/in-just-a-week-this-kit-turns-old-houses-into-zero-energy-homes-for-free (2015)

Cool Roof Rating Council, Home and Building Owners Info: (2015)

De Cian, E., Lanzi, R., Roson.: The Impact of Temperature Change on Energy Demand: A Dynamic Panel Analysis, No. I. The Fondazione Eni Enrico Mattei Note di Lavoro Series Index, 2007E. (2007)

Defaix, P.R., van Sark, W.G.J.H.M., Worrell, E., de Visser, E.: Technical potential for photovoltaics on buildings in the EU-27. Solar. Energy. **86**, 2644–2653 (2012)

Dol, K., Haffner, M.: Housing Statistics in the European Union. Ministry of the Interior and Kingdom Relations, The Hague (2010.) September
Dousset, B., Gourmelon, F., Laaidi, K., Zeghnoun, A., Giraudet, E., Bretin, P., et al.: Satellite monitoring of summer heat waves in the Paris metropolitan area. Int. J. Climatol. **31**, 313 (2011.) natio
Enterprise Community Partners, Inc: Bringing Home the Benefits of Energy Efficiency to Low-Income Households. (2008)
EU-25 Energy and Transport Outlook to 2030: Available through http://ec.europa.eu/dgs/energy_transport/figures/trends_2030/5_chap4_en.pdf (2015)
EURO4M: 2014 warmest year on record in Europe. Available through: http://cib.knmi.nl/mediawiki/index.php/2014_warmest_year_on_record_in_Europe (2015)
European Commision: Energy Renovation: The Trump Card for the New Start for Europe, JRC Science and Policy Reports (2015)
European Foundation for the Improvement of Living and Working Conditions: First European Quality of Life Survey. European Foundation for the Improvement of Living and Working Conditions, Dublin (2003)
Fanchiotti, A., Zinzi, M.: Impact of cool materials on urban heat islands and on buildings comfort and energy consumption. In: Proc. ASES Conference, 2012 (2012)
Fintikakis, N., Gaitani, N., Santamouris, M., Assimakopoulos, M., Assimakopoulos, D.N., Fintikaki, M., Albanis, G., Papadimitriou, K., Chryssochoides, E., Katopodi, K., Doumas, P: Bioclimatic design of open public spaces in the historic Centre of Tirana, Albania Original Research Article Sustainable Cities and Society. **1**(1):54–62 (2011)
Gaitani, N., Spanou, A., Saliari, M., Synnefa, A., Vassilakopoulou, K., Papadopoulou, K., Pavlou, K., Santamouris, M., Papaioannou, M., Lagoudaki, A.: Improving the microclimate in urban areas. A case study in the centre of Athens. J Building Serv Eng. **32**(1), 53–71 (2011)
GEA: Global Energy Assessment – Toward a Sustainable Future. Cambridge University Press, Cambridge, UK and the International Institute for Applied Systems Analysis, Laxenburg (2012)
Geros, V., Santamouris, M., Tsangrassoulis, A., Guarracino, G.: Experimental evaluation of night ventilation phenomena. J. Energy Buildings. **29**, 141–154 (1999)
Global Construction Perspectives and Oxford Economics: Global Construction 2025 (2013)
Hamachi, L.K., Eto, J.H.: Cost of Power Interruptions to Electricity Consumers in the United States (U.S.). Ernest Orlando Lawrence Berkeley National Laboratory (2006)
Hamilton, I., Michael Davies, a., Steadman, P., Stone, A., Ridley, I., Evans, S.: The significance of the anthropogenic heat emissions of London's buildings: A comparison against captured shortwave solar radiation. Build. Environ. **44**, 807–817 (2009)
Harvey, L.D.D.: Energy and the New Reality 1, Energy Efficiency and the Demand for Energy Services. Earthscan, London (2010)
Hassid, S., Santamouris, M., Papanikolaou, N., Linardi, A., Klitsikas, N., Georgakis, C., Assimakopoulos, D.N.: The effect of the Athens heat island on air conditioning load. Energy Build. **32**(2), 131–141 (2000)
Hermelink, A.: SOLANOVA. In: Proc European Conference and Cooperation Exchange. (2006)
House of Commons UK Parliament: Environment, Food and Rural Affairs Committee. Energy Efficiency and Fuel Poverty, Third Report of Session, 2008–09. (2009)
Hutchinson, E.J., Wilkinson, P., Hong, S.H., Oreszczyn, T., the Warm Front Study Group: Can we improve the identification of cold homes for targeted home energy-efficiency Improvements? Appl. Energy. **83**, 1198–1209 (2006)
IIASA: Energy End-Use: Buildings: Available through: http://www.iiasa.ac.at/web/home/research/Flagship-Projects/Global-Energy-Assessment/GEA_CHapter10_buildings_lowres.pdf (2015)
International Energy Agency: World Energy Investment Outlook. (2014)
Jenkings, D.P.: The value of retrofitting carbon-saving measures into fuel poor social housing. Energy Policy. **38**, 832–839 (2010)
Joint Research Center: Energy Renovation: The Trump Card for the New Start for Europe. European Commission Joint Research Centre, Institute for Energy and Transport (2015)

Juan-Carlos, C., Dowling, P.: Integrated assessment of climate impacts and adaptation in the energy sector. Energy Econ. **46**, 531–538 (2014)

Kapsomenakis, J., Kolokotsa, D., Nikolaou, T., Santamouris, M., Zerefos, S.C.: Forty years increase of the air ambient temperature in Greece: the impact on buildings. Energy Convers. Manag. **74**, 353–365 (2013)

Keatinge, W.R., Donaldson, G.C., Cordioloi, E., Martinelli, M., Kunst, A.E., Mackenbach, J.P., Nayha, S.: Heat-related mortality in warm and cold regions of Europe: observational study. Br. Med. J. **321**, 670–673 (2000)

Kolokotroni, M., Zhang, Y., Watkins, R.: The London heat island and building cooling design. Solar. Energy. **81**, 102–110 (2007)

Kolokotsa, D., Santamouris, M.: Review of the indoor environmental quality and energy consumption studies for low income households in Europe. Sci. Total Environ. **536**, 316–330 (2015)

Kolokotsa, D., Santamouris, M., Akbari, H.: Advances in the Development of Cool Materials for the Built Environment. Bentham Books, New York (2013)

Li, D., Bou-Zeid, E.: Synergistic interactions between urban heat islands and heat waves: the impact in cities is larger than the sum of its parts. J. Appl. Meteorol. Climatol. **52**, 2051–2064 (2013)

Marmot Review: The Health Impacts of Cold Homes and Fuel Poverty. http://www.foe.co.uk/sites/default/files/downloads/ cold_homes_health.pdf (2011)

Mastrapostoli, E., Karlessi, T., Pantazaras, A., Gobakis, K., Kolokotsa, D., Santamouris, M.: On the cooling potential of cool roofs in cold climates: use of cool fluorocarbon coatings to enhance the optical properties and the energy performance of industrial buildings. Energy Build. **69**, 417–425 (2014)

McDonell, G.: Displacement ventilation. Can. Archit. **48**(4), 32–33 (2003)

McGregor, G., Pelling, M., Wolf, T., Gosling, S.: The Social Impacts of Heat Waves, Science Report – SC20061/SR6. Environment Agency, Bristol (2007)

Middelkoop, B.J., Struben, H.W., Burger, I., Vroom-Jongerden, J.M.: Urban cause-specific socio-economic mortality differences. Which causes of death contribute most? Int. J. Epidemiol. **30**, 240–247 (2001)

Navigant Research: Zero Energy Buildings. Description available through http://www.navigantresearch.com/research/zero-energy-buildings (2015)

Nichol, J., King, B., Quattrochi, D., Dowman, I., Ehlers, M., Ding, X.: EO for urban planning and management state of the art and recommendations for application of earth observation in urban planning. Photogramm. Eng. Remote Sens. **73**, 973 (2007.) ote S

Odyssee-Mure: Energy Efficiency Trends in Buildings in the EU Lessons from the ODYSSEE MURE project. Available through: http://www.odyssee-mure.eu/publications/br/Buildings-brochure-2012.pdf (2012)

Pike Research: Zero Energy Buildings. Summary available through. http://www.navigantresearch.com/newsroom/revenue-from-net-zero-energy-buildings-to-reach-1-3-trillion-by-2035 (2012)

Polydoros, A., Cartalis, C.: Use of Earth observation based indices for the monitoring of built-up area features and dynamics in support of urban energy studies. Energy Build. (2014). doi:10.1016/j.enbuild.2014.09.060

Preston, I., Moore, R., Guertler, P.: How Much? The Cost of Alleviating Fuel Poverty, Report to the EAGA Partnership Charitable Trust. CSE, Bristol (2008)

Rigo, G., Parlow, E.: Modelling the ground heat flux of an urban area using remote sensing data. Theor. Appl. Climatol. **90**, 185oreti (2007)

Rübbelke, D., Vögele, S.: Distributional Consequences of Climate Change Impacts on the Power Sector: Who gains and who loses? CEPS Working Document, No. 349 (2011)

Santamouris, M. (ed.): Energy and Climate in the Urban Built Environment. Published by Earthscan, London (2001)

Santamouris, M.: On the energy impact of urban heat island and global warming on buildings. Energy Build. **82**, 100–113 (2014a)

Santamouris, M.: Cooling the cities – a review of reflective and green roof mitigation technologies to fight heat island an improve comfort in urban environments. Solar. Energy. **103**(682–703), 2014 (2014b)

Santamouris, M.: Regulating the damaged thermostat of the cities – status. Impacts. Mitig. Strateg. Energy Build. **91**, 43–56 (2015a)

Santamouris, M.: Analyzing the heat island magnitude and characteristics in one hundred Asian and Australian cities and regions. Sci. Total Environ. **512–513**, 582–598 (2015b)

Santamouris, M., Kolokotsa, D.: Passive cooling dissipation techniques for buildings and other structures: the state of the art. Energy Build. **57**, 74–94 (2013)

Santamouris, M., Kolokotsa, D.: On the impact of urban overheating and extreme climatic conditions on housing energy comfort and environmental quality of vulnerable population in Europe. Energy Build. **98**, 125–133 (2015)

Santamouris, M., Kolokotsa, D.: Urban Climate Mitigation Techniques. Earthscan, London (2016)

Santamouris, M., Papanikolaou, N., Livada, I., Koronakis, I., Georgakis, C., Argiriou, A., Assimakopoulos, D.N.: On the impact of urban climate to the energy consumption of buildings. Sol. Energy. **70**(3), 201–216 (2001)

Santamouris, M., Paraponiaris, K., Mihalakakou, G.: Estimating the ecological footprint of the heat island effect over Athens, Greece. Clim. Chang. **80**, 265–276 (2007)

Santamouris, M., Synnefa, A., Kolokotsa, D., Dimitriou, V., Apostolakis, K.: Passive cooling of the built environment – use of innovative reflective materials to fight heat island and decrease cooling needs. Int. J. Low Carbon Technol. **3**(2), 71–82 (2008)

Santamouris, M., Synnefa, A., Karlessi, T.: Using advanced cool materials in the urban built environment to mitigate heat islands and improve thermal comfort conditions. Solar. Energy. **85**, 3085–3102 (2011)

Santamouris, M., Gaitani, N., Spanou, A., Saliari, M., Gianopoulou, K., Vasilakopoulou, K.: Using cool paving materials to improve microclimate of urban areas – design realisation and results of the flisvos project. Build. Environ. **53**, 128–136 (2012a)

Santamouris, M., Xirafi, F., Gaitani, N., Spanou, A., Saliari, M., Vassilakopoulou, K.: Improving the microclimate in a dense urban area using experimental and theoretical techniques. – the case of Marousi, Athens. Int. J. Vent. **11**(1), 1–16 (2012b)

Santamouris, M.: Using cool pavements as a mitigation strategy to fight urban heat island—a review of the actual developments. Renew. Sust. Energy Rev. **26**, 224–240 (2013)

Santamouris, M., Alevizos, S.M., Aslanoglou, L., Mantzios, D., Milonas, P., Sarelli, I., Karatasou, S., Cartalis, K., Paravantis, J.A.: Freezing the poor—indoor environmental quality in low and very low income households during the winter period in Athens. Energy Build. **70**, 61–70 (2014)

Santamouris, M., Cartalis, C., Synnefa, A., Kolokotsa, D.: On the impact of urban heat island and global warming on the power demand and electricity consumption of buildings–a review. Energy Build. **98**, 119–124 (2015)

Sarrat, C., Lemonsu, A., Masson, V., Guedalia, D.: Impact of urban heat island on regional atmospheric pollution. Atmos. Environ. **40**, 1743–1758 (2006)

Sarwant, S.: New Mega Trends: Implications for Our Future Lives. Palgrave Macmillan, London (2012)

Schiano-Phan, R., Weber, F., Santamouris, M.: The mitigative potential of urban environments and their microclimates. Build. **5**, 783–801 (2015). doi:10.3390/buildings5030783

Smith, M., Whitelegg, J., Williams, N.: Greening the Built Environment. Earthscan Publications Ltd, London (1998)

Stathopoulou, M., Cartalis, C.: Use of satellite remote sensing in support of urban heat island studies. Adv. Build. Energy Res. **1**, 203–212 (2007)

Stathopoulou, E., Mihalakakou, G., Santamouris, M., Bagiorgas, H.S.: Impact of temperature on tropospheric ozone concentration levels in urban environments. J. Earth Syst. Sci. **117**(3), 227–236 (2008)

Synnefa, A., Santamouris, M.: Advances on technical, policy and market aspects of cool roof technology in Europe: the Cool Roofs project. Energy Build. **55**, 35–41 (2012)

Synnefa, A., Santamouris, M., Akbari, H.: Estimating the effect of using cool coatings on energy loads and thermal comfort in residential buildings in various climatic conditions. Energy Build. **39**(11), 1167–1174 (2007)

Synnefa, A., Karlessi, T., Gaitani, N., Santamouris, M., Assimakopoulos, D.N., Papakatsikas, C.: On the optical and thermal performance of cool colored thin layer asphalt used to improve urban microclimate and reduce the energy consumption of buildings. Build. Environ. **46**(1), 38–44 (2011)

Synnefa, A., Saliari, M., Santamouris, M.: Experimental and numerical assessment of the impact of increased roof reflectance on a school building in Athens. Energy Build. **55**, 7–15 (2012)

Thiers, S., Peuportier, B.: Thermal and environmental assessment of a passive building equipped with an earth-to-air heat exchanger in France. Solar Energy. **82**(9), 820–831 (2008)

United Nations: Urban and Rural Areas.: www.unpopulation.org (2009)

Urge-Vorsatz, D., Eyre, N., Graham, P., Harvey, D., Hertwich, E., Kornevall, C., Majumdar, M., McMahon, J., Mirasgedis, S., Murakami, S., Novikova, A., Yi, J.: Energy end-use: buildings. In: The Global Energy Assessment: Toward a More Sustainable Future. IIASA, Laxenburg and Cambridge University Press, Cambridge, UK (2012)

U.S. Department of Energy: Weatherization Assistance Program: Improving the Economies for Low-Income Communities. (2006)

Van Vliet, M.T.H., Yearsley, J.R., Ludwig, F., Vögele, S., Lettenmaier, D.P., Kabat, P.: Vulnerability of US and European electricity supply to climate change. Nat. Clim. Chang. **2**, 676–681 (2012)

Vision Gain: Flat Glass Market Report 2014–2024, Opportunities For Leading Companies in Building, Construction, Automotive & Solar Energy. (2014) https://www.visiongain.com/report_license.aspx?rid=1337

Washan, P., Stenning, J., Goodman, M.: Building the Future: Economic and Fiscal Impacts of Making Homes Energy Efficient. Cambridge Econometrics (2014)

WIEGO: Woman in Informal Economy: Globalizing and Organising: Construction Workers. Available through http://wiego.org/informal-economy/occupational-groups/construction-workers (2015)

Xu, W., Wooster, M.J., Grimmond, C.S.B.: Modelling of urban sensible heat flux at multiple spatial scales: a demonstration using airborne hyperspectral imagery of Shanghai and a temperatureple spatial scales: a demonstrat. Remote. Sens. Environ. **112**, 3493 Sensi (2008)

Yohanis, Y.G., Mondol, J.D.: Annual variation of temperature in a sample of UK dwellings. Appl. Energy. **87**(2), 681–690 (2010)

Zerohomes: Zero Energy Retrofits Benefits of Remodeling Existing Homes to Zero or Near Zero. Available through http://www.zerohomes.org/zero-energy-retrofitting/ (2015)

Systems Science Simulation Modeling to Inform Urban Health Policy and Planning

Yan Li, Jo Ivey Boufford, and José A. Pagán

Abstract More than half of the population in the world lives in cities and urban populations are still rapidly expanding. Increasing population growth in cities inevitably brings about the intensification of urban health problems. The multidimensional nature of factors associated with health together with the dynamic, interconnected environment of cities moderates the effects of policies and interventions that are designed to improve population health. With the emergence of the "Internet of Things" and the availability of "Big Data," policymakers and practitioners are in need of a new set of analytical tools to comprehensively understand the social, behavioral, and environmental factors that shape population health in cities. Systems science, an interdisciplinary field that draws concepts, theories, and evidence from fields such as computer science, engineering, social planning, economics, psychology, and epidemiology, has shown promise in providing practical conceptual and analytical approaches that can be used to solve urban health problems. This chapter describes the level of complexity that characterizes urban health problems and provides an overview of systems science features and methods that have shown great promise to address urban health challenges. We provide two specific examples to showcase systems science thinking: one using a system dynamics model to prioritize interventions that involve multiple social determinants of health in Toronto, Canada, and the other using an agent-based model to evaluate the impact of different food policies on dietary behaviors in

Y. Li
The New York Academy of Medicine, 1216 Fifth Avenue, New York, NY, 10029, USA

Department of Population Health Science and Policy, Icahn School of Medicine at Mount Sinai, New York, NY, USA

J.I. Boufford
The New York Academy of Medicine, 1216 Fifth Avenue, New York, NY, 10029, USA

J.A. Pagán (✉)
Center for Health Innovation, The New York Academy of Medicine, 1216 Fifth Avenue, New York, NY, 10029, USA

Department of Public Health Policy and Management, College of Global Public Health, New York University, New York, NY, USA

Leonard Davis Institute of Health Economics, University of Pennsylvania, Philadelphia, PA, USA
e-mail: jpagan@nyam.org

S.T. Rassia, P.M. Pardalos (eds.), *Smart City Networks*, Springer Optimization and Its Applications 125, DOI 10.1007/978-3-319-61313-0_9

New York City. These examples suggest that systems science has the potential to foster collaboration among researchers, practitioners, and policymakers from different disciplines to evaluate interconnected data and address challenging urban health problems.

1 Introduction

Fifty-four percent of the population in the world lived in cities in 2015, and this number is expected to exceed 60% by 2030 (World Health Organization 2016). The rapid growth of urban populations across the globe has resulted in unprecedented pressure to the general infrastructure of cities and communities. If such growth is poorly managed, it can have a negative impact in many areas of urban life—such as air quality, housing, and transportation—and eventually may have a detrimental effect on the health of urban residents (Vlahov et al. 2011). However, if cities are managed well, there is an opportunity for decisions in all sectors to be assessed for their positive and negative effects on health, and the environment in cities can be positive. Improving the health of urban residents requires policymakers to not only think about ways to improve healthcare services delivery but also come up with innovative solutions to promote disease prevention and address social determinants of health (Marmot and Wilkinson 2005; Thomas et al. 2016). This could be a daunting task for many health policymakers, especially in this era when massive data is available and multi-sectoral social determinants are often connected due to recent advancements in information technology and the Internet of Things (Atzori et al. 2010). As a result, many urban health policies fail to deliver on their intended outcomes or improve population health in a sustained manner (Corburn 2004).

Urban environments are often characterized or described as dense, diverse, dynamic, and complex (Vlahov et al. 2007), and, within each feature, there are possibilities for positive and negative effects for policymakers to consider as they develop strategies and solutions to improve urban health. For example, the increasing density of urban populations, if unmanaged, can facilitate the rapid spread of infectious diseases, but, if appropriately managed, the density of populations can facilitate access to critical health and social services. The diversity of urban populations indicates that interventions that are effective for a specific population may not work well for another group due to social and cultural differences across different populations, but this diversity can also create rich local cultures that can support cohesiveness within communities.

Cities are also dynamic in many aspects—migration, the construction of new housing and other infrastructure, changes in public transportation, and air pollution dynamics—all of which are constantly affecting the health of urban residents. This dynamism also makes them hubs of innovation and economic development. Urban health problems are also complex; a health epidemic often involves multiple factors at different levels across different sectors and, thus, simple interventions can rarely be sufficient to solve a given problem.

High levels of health disparities often exist across different social groups and different neighborhoods in urban areas. These disparities may be further reinforced by inequities in social determinants of health (e.g., local food environment, housing, education and economic development, etc.) and healthcare resources within and across neighborhoods. It is important that interventions designed to improve population health should also seek to reduce disparities as an important additional aim; otherwise, population-based strategies may have the unintended effect of widening existing disparities. For example, a recent study shows that screening and treating individuals at high risk for cardiovascular disease (CVD) at the population level may widen disparities (Capewell and Graham 2010). In contrast, multi-sectoral, population-wide interventions (e.g., taxation of sugar sweetened beverages, subsidizing fruits and vegetables) that focus on disease prevention may be more effective in reducing health disparities (Kypridemos et al. 2016; Fawcett et al. 2010).

To fully understand the density, diversity, dynamism, and complexity of urban health problems and reduce health disparities, systems thinking and system-level analytical tools are required. This chapter aims to articulate the rationale for using systems science models to solve urban health problems and describes the two most commonly used modeling techniques—system dynamics modeling and agent-based modeling. An example of each modeling approach in addressing challenging urban health problems is then presented. Barriers in applying systems science models to inform urban health policies are identified. The chapter concludes with a summary of important future research directions.

2 Rationale for a Systems Science Approach

Systems science is an interdisciplinary area of work focused on the study of complex systems which are usually characterized by heterogeneous entities interacting with each other, stochastic and dynamic emergent effects, and adaptive processes (Sawyer 2005). While traditional epidemiologic approaches can identify important associations and causal mechanisms between exposures and outcomes from a reductionist point of view, they are limited in capturing interdependent, dynamic relationships, feedback loops, and population interactions that prevail among many urban health problems. A systems approach, instead, can complement traditional epidemiologic approaches by taking a holistic view of challenging urban health problems, exploring the underlying mechanisms that produce undesirable population health outcomes and health disparities, and seeking innovative solutions that align with the social and cultural traits of the population (Luke and Stamatakis 2012; Mabry et al. 2013).

Another reason for using a systems science approach to tackle urban health problems is that cities essentially consist of networks and flows, and, thus, a deep understanding of urban problems—including urban health problems—requires tools from complex systems (Batty 2013; Forrester and Forrester 1969). Urban health researchers must recognize the evolution of health problems within the environment in cities as an interconnected system rather than a set of isolated components; the effect of an intervention on an urban health problem depends on the relationship of different city components. For example, insufficient physical activity and the growing prevalence of obesity in a neighborhood may be associated with a poorly designed built environment and/or a low walkability of a given location (Handy et al. 2002). In addition, promoting healthy eating may require an in-depth understanding of the socioeconomic and cultural factors behind dietary choices, and, consequently, interventions that leverage social networks and target social norms may prove effective in an urban setting (Zhang et al. 2014).

Over the past few decades, the health concerns for urban residents have increasingly shifted from traditional infectious disease and harmful environmental exposures to chronic health conditions such as obesity, mental disorders, hypertension, and diabetes (Diez Roux 2015). These chronic health conditions, often linked to sedentary lifestyles, poor diets, poverty, and even violence, are very difficult to address in urban settings. Specifically, these chronic health conditions present the following challenges to public health practitioners and policymakers: (1) their development involves a range of factors at multiple levels (e.g., biological, behavioral, social, and environmental factors), (2) they present heterogeneous characteristics across different populations and geographic locations (e.g., an intervention that proved effective for a population may be less effective for a different population), (3) there exist complex interactions among populations (e.g., members in a social network influencing the health attitudes and preferences of others), and (4) the impact of some prevention programs (e.g., smoking cessation, nutrition education) can only be seen in the long term, which makes it difficult for program evaluation and the short-term assessment of outcomes. The challenges to curb the chronic disease epidemic in urban settings further justify the use of a systems science approach, which provides a "big picture" and perhaps even a coherent view of the evolution of health and the consequences of chronic diseases and identifies important drivers of urban health problems and health disparities across multiple sectors (Diez Roux 2011).

Finally, designing and developing a valid, practical systems science model require constant communication and close collaboration within a multidisciplinary team structure. The team of experts must include systems science modelers as well as public health researchers and practitioners, epidemiologists, urban health experts, and city planners (and many other experts depending on the problem at hand). The model development process should not only identify the elements and factors that are fundamental to understand the target problem and formulate it in a simulation framework but also identify current gaps in data collection, explore new research questions, and draw new insights into old problems. Thus, a systems science approach also presents a potentially effective way to facilitate interdisciplinary communication and collaboration while also providing a platform for different stakeholders in cities to come up with creative ideas to address urban health problems.

3 System Dynamics and Agent-Based Modeling

In this chapter we focus on two popular systems science approaches—system dynamics and agent-based modeling. Other approaches, such as network analysis and discrete-event simulation, have also been used in public health research in recent years (Fabian et al. 2012; Luke and Stamatakis 2012). The choice of the modeling approach usually depends on the nature of the research question and the type of data available, as well as the familiarity of each modeling method within the research team.

3.1 System Dynamics

System dynamics is a simulation modeling approach that can be used to better understand complex, nonlinear relationships of a system to then understand behaviors over time. A system dynamics model aggregates individuals into homogeneous groups (stocks) and captures changes of population health profiles by simulating transitions (flows) among different groups. The modeling approach excels in studying complex feedback loops (bidirectional relationships among risk factors and health outcomes) in population health. A model based on system dynamics operates in continuous time and often uses differential equations to explain the relationship between different factors and entities.

System dynamics models typically require system-level data to capture the dynamics of the system, and the modeling approach has the capacity to combine qualitative data with quantitative data to better understand data patterns or predict outcomes over time (Homer and Hirsch 2006). System dynamics models are useful for recognizing behavioral patterns in a system, gaining insight into the processes of a system, and identifying leverage points for system redesign to reproduce a given behavior.

System dynamics models are particularly useful for environmental and public health planning such as evaluating urban policies to reduce CO_2 emissions, informing healthcare policy related to screening and preventive practices, and designing disease interventions in large healthcare systems (Fong et al. 2009; Milstein et al. 2007, 2010). The most popular commercial software readily available for developing system dynamics models include AnyLogic (The AnyLogic Company, St. Petersburg, Russian Federation), iThink/Stella (ISEE Systems, Lebanon, New Hampshire, USA), Powersim (Powersim Software, Nyborg, Norway), and Vensim (Ventana Systems, Cambridge, Massachusetts, USA).

3.2 Agent-Based Modeling

Agent-based modeling studies system-level emergent phenomena through explicitly simulating individual health and behaviors (Epstein 2006). Agent-based models are flexible in capturing real-world phenomena because agents can be endowed with a large set of properties such as the ability to interact, heterogeneity, randomness, adaptability, and mobility (Bonabeau 2002). Compared to system dynamics, agent-based modeling is usually used to solve problems in which individual-level behaviors and interactions are relevant and data on heterogeneous characteristics of individuals are available (Macal and North 2010).

In public health and medicine, agent-based models have been mostly used to simulate the epidemics of infectious diseases such as influenza and sexually transmitted diseases (Eubank et al. 2004; Kumar et al. 2013). By generating populations of different characteristics and incorporating rules that govern disease transmission, agent-based models can capture the entire course of a disease outbreak and evaluate the impact of alternative interventions at local or global levels.

Recently, agent-based models have been increasingly used to study chronic health conditions and inform public health policies related to, for example, diabetes and cardiovascular disease. Two literature reviews are available to provide an overview of agent-based models of chronic health conditions and health behaviors (Li et al. 2016a; Nianogo and Arah 2015). There is also a growing number of agent-based modeling software packages that can greatly facilitate the model development process and make modeling and results more accessible to health researchers—the most popular being AnyLogic (The AnyLogic Company, St. Petersburg, Russian Federation), NetLogo (The Center for Connected Learning and Computer-Based Modeling, Northwestern University, Chicago, Illinois, USA), Repast (Argonne National Laboratory, Lemont, Illinois, USA), and Swarm (Swarm Development Group, Santa Fe, New Mexico, USA).

4 Examples of Systems Science Models to Improve Urban Health

Although the application of systems science models is not rare in public health research, few models have been explicitly designed to target urban health problems by incorporating properties such as the density, diversity, dynamism, and complexity of cities. To show the promise of systems science to facilitate multi-sectoral collaborations, improve urban health, and reduce health inequities, we present two examples of urban health systems science models, based on real data and problems for the cities of Toronto and New York.

4.1 A System Dynamics Model of Social Determinants in Toronto

Numerous studies have shown that population health and well-being are greatly affected by social determinants such as education, income, access to care, and even racial discrimination (Braveman et al. 2011). Although knowledge on how particular social determinants are associated with health outcomes is growing, the exact mechanism through which social determinants interact and shape health outcomes across different populations is still unclear. In urban settings, it is particularly challenging to understand the relationships between social determinants and health outcomes and leverage this knowledge to design effective population health interventions given limited resources in these complex and dynamic environments (Tozan and Ompad 2015).

Mahamoud et al. (2013) developed a system dynamics model of social determinants and health outcomes based on population data from the City of Toronto, Canada. The modeling team engaged key stakeholders in the city—including community residents, representatives from nongovernment and community-based organizations, and policymakers at the municipal level—and used a participatory strategy to identify a range of urban health issues such as the high prevalence of chronic health conditions and health disparities. Stakeholders provided important qualitative (e.g., opinions, observations) and quantitative (e.g., specific datasets, estimates of model parameters) information that was crucial to model development. The modeling team went through an iterative process to refine the model structure and parameters based on feedback from stakeholders and, eventually, used the model as a virtual policy laboratory to study the impact of alternative policy options on health outcomes for as long as 40 years into the future.

The model developed by Mahamoud et al. (2013) is able to track the health progression of 30 different population groups stratified by gender (male or female), ethnicity (Black, East Asian, Southwest Asian, White, or other), and immigration status (recent, established, native born) for adults 25–64 years of age. Key factors in the model include smoking, obesity, chronic illness, access to healthcare, disability, housing, social cohesion, low-income, and mortality. Potential areas of interventions include improving healthcare access, promoting healthy behaviors, increasing income, improving housing, and enhancing social cohesion. The health outcomes in the model include changes in mortality rates, the prevalence of health and social determinants, and ratios capturing changes in health disparities. Figure 1 provides an overview of the model structure with causal pathways among social determinants and health outcomes. The model is parameterized with data specific to the City of Toronto from sources such as the Canadian Community Health Survey and the Canadian Census.

In the simulation experiments, the authors compared the impact of five hypothetical interventions—improving housing, increasing healthcare access, strengthening social cohesion (a sense of community belonging), reducing unhealthy behaviors, and reducing the low-income population by 30%, respectively—on the number

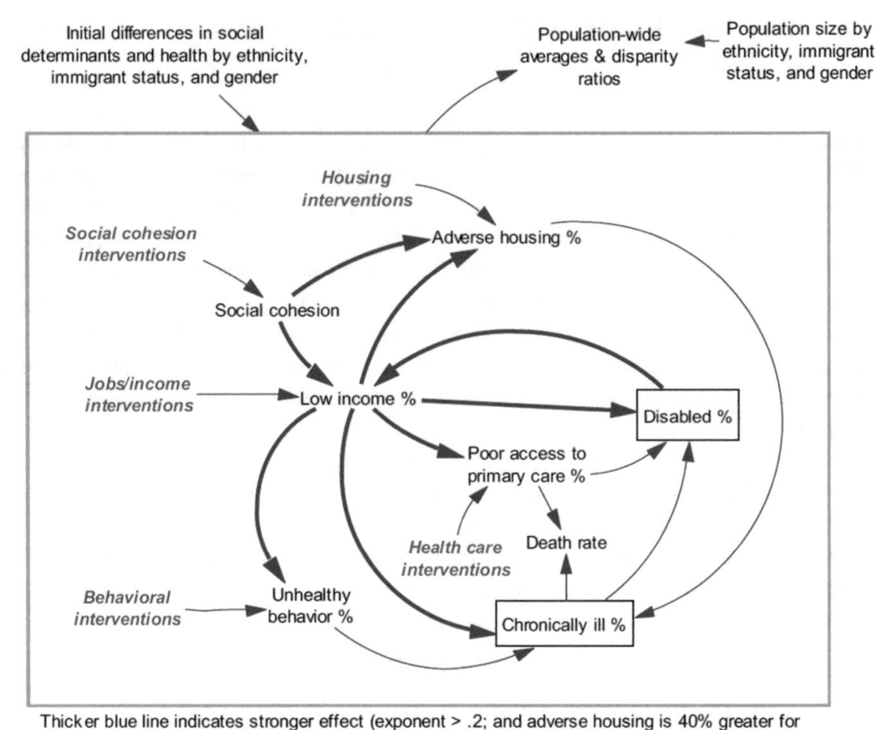

Fig. 1 Overview of the system dynamics model of social determinants of health for the City of Toronto (Mahamoud et al. 2013)

of disabled adults 25–64 years of age in the city. They found that, among the five interventions, strengthening social cohesion and closing income gaps would have the most significant impact in reducing the size of the disabled population in 40 years (Fig. 2). They also tested single interventions with varying magnitudes of effectiveness and combined interventions, which resulted in valuable policy insights that could be taken into account to improve the health outcomes of people in the City of Toronto. For example, they found that the effects of income and social cohesion interventions are evident early within the first simulated decade, while the effects of improving unhealthy behaviors are evident in later decades. These insights may be useful to help urban health policymakers decide the timing of interventions based on short-term versus long-term objectives. In addition, they found that a combined intervention consisting of 30% improvements in adverse housing and social cohesion would significantly reduce the prevalence of chronic illness in 40 years. Combined interventions such as this one—which lever the interactions and feedback effects of social determinants and health outcomes—may be more effective in tackling some of the most challenging urban health problems.

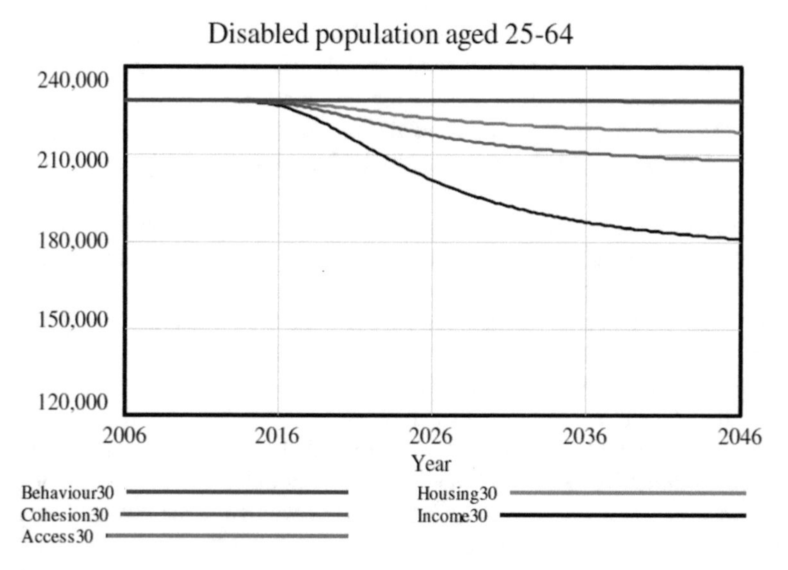

Fig. 2 Comparison of five intervention scenarios based on the system dynamics model of social determinants of health for the City of Toronto (Mahamoud et al. 2013)

4.2 An Agent-Based Model of Dietary Behaviors in New York City

Consumption of fruit and vegetables is associated with a lower risk of obesity, diabetes, hypertension, and cardiovascular disease. However, only about 10% of New York City residents consume the recommended amount (five or more servings) of fruits and vegetables each day (Jack et al. 2013). Low consumption of fruits and vegetables is a long-standing urban health problem in New York City and accounts for the high prevalence of many chronic conditions and significant health disparities. It is also a complex problem because an individual's dietary choice can be influenced by many interdependent, dynamic factors such as income, taste preferences, the local food environment, and even the influence of family and peers.

To fully understand how these different factors may be related to the dietary behaviors of New Yorkers and inform public health policy, Li et al. (2016b) developed an agent-based model of dietary behaviors for residents in New York City. In their model, an agent can be either an individual or a food outlet. Individuals differ by their demographic characteristics (i.e., age, gender, and educational attainment) and cognitive habits (i.e., taste preferences, health beliefs, price sensitivity). Food outlets include healthy outlets (supermarkets and fruit and vegetable markets) and less healthy outlets (fast-food restaurants). The geographic location of individuals and food outlets is determined by the size and layout of the simulated neighborhood. In the model, each neighborhood is represented by a grid of cells—each cell measuring $26.4 \times 26.4\text{ft}^2$ (0.005×0.005 miles^2). One time period in the model represents 1 day, and each simulation runs for 3 years (1095 days).

Each individual makes daily dietary decisions based on his/her health beliefs, taste preferences, price sensitivity, food accessibility, and basic demographic characteristics. Data from the 2007 Food Attitudes and Behaviors (FAB) survey were used to construct a probit equation of the probability of consuming at least two servings of fruits and vegetables a day. Individuals are connected in a small-world social network based on the Watts-Strogatz framework (Watts and Strogatz 1998). Simulated individuals belong to groups of friends within their community, and they also may have friendship ties to members of other groups as well. The taste preferences and health beliefs of a simulated individual may change over time in response to the social norms formed by peers in the social network (Zimmerman 2013).

This agent-based model went through calibration and validation exercises, including predictive validation. As part of model validation, Table 1 shows that the simulated results are close to actual estimated values for the consumption of fruits and vegetables across different New York City neighborhoods obtained from the Community Health Survey. Thus, the model could be useful for public health researchers and policymakers interested in evaluating the potential impact of different interventions and policies designed to improve the consumption of fruits and vegetables across different neighborhoods in New York City. Examples of these interventions and policies include the evaluation of existing or new regulations on the local food environment, the effect of taxes on the consumption of unhealthy food staples, the impact of healthy food subsidies, and the influence of mass media campaigns or nutrition education programs to promote healthy eating behaviors (see, e.g., related work on healthy food consumption in downtown Los Angeles (Zhang et al. 2014).

The Li et al. (2016a) study simulated two hypothetical mass media campaign and nutrition education interventions that could improve the influence of positive social norms (a person's knowledge of what others do) by 5% and 10%, respectively, and predicted the impact of these interventions on the proportion of the population who consume two or more servings of fruits and vegetables across 34 New York City neighborhoods (defined by the United Hospital Fund) in 3 years. The study found that a 5% improvement on positive social norms was associated with an increase in the proportion of the population consuming two or more servings of fruits and vegetables that ranged between 0.58% and 8.97% across the 34 neighborhoods considered. A 10% improvement on positive social norms was related to an increase in the population consuming two or more servings of fruits and vegetables that ranged between 2.68% and 13.94% across the same 34 neighborhoods. Figure 3 presents a New York City map demonstrating the consumption of fruits and vegetables before and after the simulated intervention in each neighborhood.

Further analysis of the simulation results revealed that the impact of interventions that work on the effect of peer influence may be less effective in neighborhoods with relatively low education levels. It is possible that residents in these neighborhoods have other competing priorities—such as housing and safety—that may result in poorer food consumption choices. In addition, other factors such as a relatively high sensitivity to food prices or poor access to healthy food establishments may

Table 1 Proportion of the population consuming at least two servings of fruits and vegetables per day by New York City neighborhood (comparison between NYC CHS and model simulation)

UHF code	Name	CHS (%)	Simulation (%)
101	Kingsbridge – Riverdale	66.90	68.13
102	Northeast Bronx	62.90	62.96
103	Fordham – Bronx Park	56.42	56.47
104	Pelham – Throgs Neck	63.37	64.60
105/106/107	South Bronx	53.56	54.74
201	Greenpoint	73.58	73.17
202	Downtown – Heights – Slope	72.48	72.36
203	Bedford Stuyvesant – Crown Heights	54.65	57.65
204	East New York	58.78	59.03
205	Sunset Park	69.02	60.49
206	Borough Park	72.27	73.04
207	East Flatbush – Flatbush	61.99	61.09
208	Canarsie – Flatlands	67.52	68.39
209	Bensonhurst – Bay Ridge	68.97	69.53
210	Coney Island – Sheepshead Bay	73.18	75.94
211	Williamsburg – Bushwick	58.84	59.55
301	Washington Heights – Inwood	60.81	59.36
302	Central Harlem – Morningside Heights	61.68	60.67
303	East Harlem	73.16	73.68
304	Upper West Side	64.62	68.23
305/307	Upper East Side – Gramercy	81.30	82.44
306/308	Chelsea – Village	78.30	80.33
309/310	Union Square, Lower Manhattan	72.55	73.06
401	Long Island City – Astoria	72.35	73.34
402	West Queens	68.54	67.83
403	Flushing – Clearview	70.55	69.91
404/406	Bayside – Meadows	74.27	74.91
405	Ridgewood – Forest Hills	75.29	75.13
407	Southwest Queens	71.37	71.45
408	Jamaica	64.74	63.15
409	Southeast Queens	61.72	62.00
410	Rockaway	59.80	59.52

These results come from the agent-based model and data analyses described in Li et al. (2016b); they had not been published previously

also result in a lower consumption of fruits and vegetables in these neighborhoods. While it is clear that residents in neighborhoods with relatively low education would need more resources to substantially change their eating behaviors, the agent-based model introduced here can be an effective tool to help policymakers and planners prioritize alternative interventions and policies in urban environments.

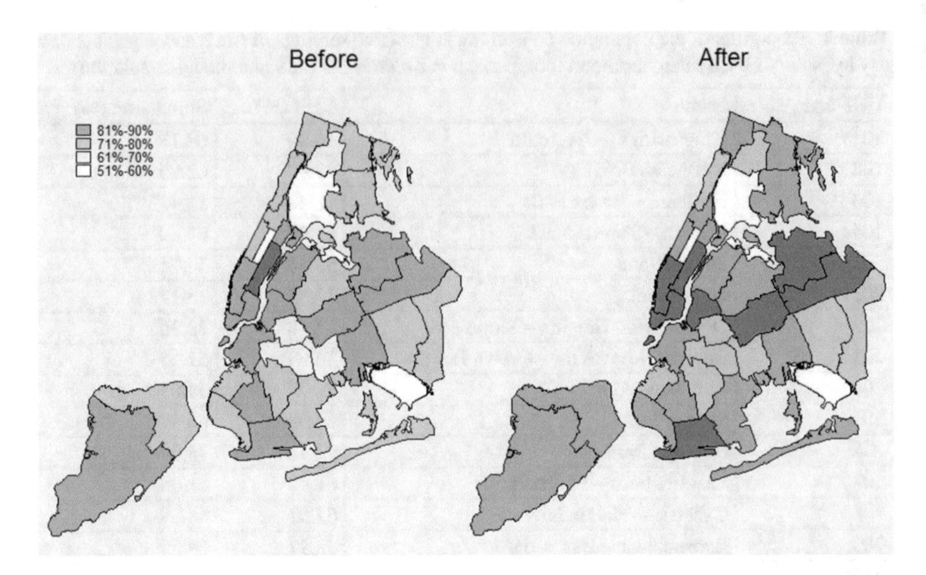

Fig. 3 Impact of a proposed mass media and nutrition education campaign on fruit and vegetable consumption in New York City – before and after policy simulation (Li et al. 2016b) (Source: Neighborhood-level consumption before the intervention was based on authors' analysis of data from the NYC Community Health Survey. Consumption estimates after the intervention were based on simulation of the proposed intervention. Notes: Comsumption of fruits and vegetables was measured by the proportion of the population who consume two or more servings of fruits and vegetables per day within a neighborhood)

5 Barriers to Implementing a Systems Science Approach

Although there is a growing interest in using systems science methodologies to solve complex public health and urban health problems, significant barriers within both research and practice communities still exist and need to be recognized in order to move forward. First, collaboration between systems science modelers, public health practitioners, urban planners, and policymakers is usually difficult to develop and sustain due to differences in training backgrounds, differing views on how to address health problems, and financial constraints. For example, systems science modelers tend to focus on methodological rigor and innovation when thinking about a specific health problem, whereas public health practitioners are typically focused on finding real-world solutions to complex problems. These differences in approaches to problem-solving and conceptualization oftentimes generate disagreement during the model development process, which must be addressed through improved and inclusive interdisciplinary communication.

Second, there are several misconceptions about the benefits and limitations of systems science modeling within the public health community, which may lead to unrealistic expectations and even dissatisfaction with what systems science can do to address complex health challenges. Systems science—like many other analytical approaches—relies on simplifying the real world and, as a result, may suffer from potential issues such as ill-defined model structure, missing or poor-quality data, and computational constraints. The value of a systems science approach is not to produce accurate predictions but to allow decision-makers to draw insights from low-risk simulated interventions about what factors, dynamics, and interactions drive the emergence of population health outcomes. That being said, a well-designed, validated model is able to make reasonable predictions of the future, with quantifiable uncertainty. The value of a systems science approach also lies in its ability of synthesizing existing evidence and data—both quantitative and qualitative whenever available—and provide decision-makers with a holistic view of a given problem or sets of problems and an in-depth understanding of complex phenomena in the real world.

Finally, there exists limited funding in systems science modeling in health or interdisciplinary research in general. Academic institutions and funding agencies have the tendency to reward field-specific expertise, while important interdisciplinary research, such as systems science modeling, has difficulty establishing and positioning itself within the existing academic research financing structure. Despite these challenges, there are some promising signs that systems science research and interdisciplinary collaboration are becoming increasingly accepted and relevant. For example, government agencies in the United States such as the National Institutes of Health and the Centers for Disease Control and Prevention have issued funding announcements encouraging the use of systems science methodologies. Philanthropic organizations such as the Robert Wood Johnson Foundation have also funded the development and application of systems science models to address chronic health conditions.

6 Conclusion

Urban health policy and planning will benefit substantially from the development of new systems science simulation models that can be used to study complex urban health problems, compare different strategies, and inform policy decisions. New models are increasingly sophisticated, and increases in computing power, new software availability, and improved data visualization tools are making model development and testing easier and more widely available. In addition, the recent development and growth of the Internet of Things—which encompasses the collection of large amounts of interconnected data between people and objects through sensors, tags, and mobile phones—opens up more opportunities for systems science modelers to use these new sources of data to better reflect the real world (Gubbi et al. 2013). Cities are characterized by density, complexity, diversity, and dynamism and are at risk of significant health disparities across communities. As the global urban population continues to grow unabated, systems science approaches will become increasingly more important for policymaking and to understand the consequences of urban growth.

Our two examples showcasing how system dynamics and agent-based modeling can be useful to address social determinants of health in urban contexts provide concrete ways in which systems science modeling can be used to engage partners in the community and compare relevant strategies and options. A key lesson learned is that systems science model building requires interdisciplinary collaboration in order for researchers to be able to build useful models that can be validated and adopted broadly. Given that systems science models can grow very quickly on their degree of complexity, one recommendation is to begin with simple, clearly articulated research questions and problem statements that allow everyone in the interdisciplinary team to contribute to the development and growth of the model in ways that keep all engaged. Constant communication and interaction between model developers and urban health experts and policymakers and engaging communities of interest are required so that model developers clearly understand the substantive health problems being studied and all the different pathways and interactions across factors critical to the model, while policymakers and community stakeholders can understand the potential as well as the limitations of systems science modeling strategies. In the end, the widespread adoption of systems science modeling in urban health policy and planning will depend on how these issues are addressed model by model, as well as which level of resources are invested by the public and private sectors to accelerate these efforts.

References

Atzori, L., Iera, A., Morabito, G.: The internet of things: a survey. Comput. Netw. **54**(15), 2787–2805 (2010)

Batty, M.: The New Science of Cities. MIT Press (2013). Retrieved from https://books.google.com/books?hl=en&lr=&id=yX-YAQAAQBAJ&oi=fnd&pg=PR7&dq=the+new+science+of+cities&ots=2jOm4VDely&sig=aPS7_YRLt-HT4QvYsV5pnhz6nvM

Bonabeau, E.: Agent-based modeling: methods and techniques for simulating human systems. Proc. Natl. Acad. Sci. **99**(suppl 3), 7280–7287 (2002)

Braveman, P.A., Egerter, S.A., Mockenhaupt, R.E.: Broadening the focus: the need to address the social determinants of health. Am. J. Prev. Med. **40**(1), S4–S18 (2011)

Capewell, S., Graham, H.: Will cardiovascular disease prevention widen health inequalities? PLoS Med. **7**(8), e1000320 (2010)

Corburn, J.: Confronting the challenges in reconnecting urban planning and public health. Am. J. Public Health. **94**(4), 541–546 (2004)

Diez Roux, A.V.: Complex systems thinking and current impasses in health disparities research. Am. J. Public Health. **101**(9), 1627–1634 (2011)

Diez Roux, A.V.: Health in cities: is a systems approach needed? Cad. Saude Publica. **31**, 9–13 (2015)

Epstein, J.M.: Generative Social Science: Studies in Agent-Based Computational Modeling. Princeton University Press, Princeton (2006)

Eubank, S., Guclu, H., Kumar, V.A., Marathe, M.V., Srinivasan, A., Toroczkai, Z., Wang, N.: Modelling disease outbreaks in realistic urban social networks. Nature. **429**(6988), 180–184 (2004)

Fabian, M.P., Stout, N.K., Adamkiewicz, G., Geggel, A., Ren, C., Sandel, M., Levy, J.I.: The effects of indoor environmental exposures on pediatric asthma: a discrete event simulation model. Environ. Health. **11**(1), 66 (2012)

Fawcett, S., Schultz, J., Watson-Thompson, J., Fox, M., Bremby, R., et al.: Building multisectoral partnerships for population health and health equity. Prev. Chronic Dis. **7**(6), A118 (2010)

Fong, W.-K., Matsumoto, H., Lun, Y.-F.: Application of system dynamics model as decision making tool in urban planning process toward stabilizing carbon dioxide emissions from cities. Build. Environ. **44**(7), 1528–1537 (2009)

Forrester, J.W., Forrester, J.W.: Urban Dynamics, vol. 114. MIT Press, Cambridge (1969). Retrieved from http://search.proquest.com/openview/111accf0f58d4e948aa8aa8e8e44530a/1.pdf?pq-origsite=gscholar&cbl=35192

Gubbi, J., Buyya, R., Marusic, S., Palaniswami, M.: Internet of Things (IoT): a vision, architectural elements, and future directions. Future Gen. Comput. Syst. **29**(7), 1645–1660 (2013)

Handy, S.L., Boarnet, M.G., Ewing, R., Killingsworth, R.E.: How the built environment affects physical activity: views from urban planning. Am. J. Prev. Med. **23**(2), 64–73 (2002)

Homer, J.B., Hirsch, G.B.: System dynamics modeling for public health: background and opportunities. Am. J. Public Health. **96**(3), 452–458 (2006)

Jack, D., Neckerman, K., Schwartz-Soicher, O., Lovasi, G.S., Quinn, J., Richards, C.: Socioeconomic status, neighborhood food environments and consumption of fruits and vegetables in New York City. Public Health Nutr. **16**(7), 1197–1205 (2013)

Kumar, S., Grefenstette, J.J., Galloway, D., Albert, S.M., Burke, D.S.: Policies to reduce influenza in the workplace: impact assessments using an agent-based model. Am. J. Public Health. **103**(8), 1406–1411 (2013)

Kypridemos, C., Allen, K., Hickey, G.L., Guzman-Castillo, M., Bandosz, P., Buchan, I., Capewell, S., O'Flaherty, M.: Cardiovascular screening to reduce the burden from cardiovascular disease: microsimulation study to quantify policy options. BMJ. **353**, i2793 (2016)

Li, Y., Lawley, M. A., Siscovick, D. S., Zhang, D., Pagán, J.A.: Agent-based modeling of chronic diseases: a narrative review and future research directions. Prev. Chronic Dis. **13**, 150561 (2016a). Retrieved from http://origin.glb.cdc.gov/pcd/issues/2016/15_0561.htm

Li, Y., Zhang, D., Pagán, J.A.: Social norms and the consumption of fruits and vegetables across New York City neighborhoods. J. Urban Health. **93**(2), 244–255 (2016b)

Luke, D.A., Stamatakis, K.A.: Systems science methods in public health: dynamics, networks, and agents. Annu. Rev. Public Health. **33**, 357–376 (2012)

Mabry, P.L., Milstein, B., Abraido-Lanza, A.F., Livingood, W.C., Allegrante, J.P.: Opening a window on systems science research in health promotion and public health. Health Educ. Behav. **40**(1 suppl), 5S–8S (2013)

Macal, C.M., North, M.J.: Tutorial on agent-based modelling and simulation. J. Simul. **4**(3), 151–162 (2010)

Mahamoud, A., Roche, B., Homer, J.: Modelling the social determinants of health and simulating short-term and long-term intervention impacts for the city of Toronto, Canada. Soc. Sci. Med. **93**, 247–255 (2013)

Marmot, M., Wilkinson, R.: Social Determinants of Health. OUP, Oxford (2005). Retrieved from https://books.google.com/books?hl=en&lr=&id=AmwiS8HZeRIC&oi=fnd&pg=PA17&dq=social+determinants+of+health+urban+health&ots=y_IBd2UvD1&sig=qAMeWsnozsUDTiJ9H7Gf8VGBoPE

Milstein, B., Jones, A., Homer, J.B., Murphy, D., Essien, J., Seville, D.: Charting plausible futures for diabetes prevalence in the United States: a role for system dynamics simulation modeling. Prev. Chronic Dis. **4**(3), A52 (2007). Retrieved from http://www.ncbi.nlm.nih.gov/pmc/articles/PMC1955415/

Milstein, B., Homer, J., Hirsch, G.: Analyzing national health reform strategies with a dynamic simulation model. Am. J. Public Health. **100**(5), 811–819 (2010)

Nianogo, R.A., Arah, O.A.: Agent-based modeling of noncommunicable diseases: a systematic review. Am. J. Public Health. **105**(3), e20–e31 (2015)

Sawyer, R.K.: Social Emergence: Societies as Complex Systems. Cambridge University Press, Cambridge (2005). Retrieved from https://books.google.com/books?hl=en&lr=&id=Hgs007Rd_moC&oi=fnd&pg=PP11&dq=,+Social+Emergence:++Societies+as+Complex+Systems&ots=IagDx3YbrC&sig=GTkC_be-BbKlMHggnYD1XXDtCtA

Thomas, Y.F., Boufford, J.I., Talukder, S.H.: Focusing on health to advance sustainable urban transitions. J. Urban Health. **93**(1), 1–5 (2016)

Tozan, Y., Ompad, D.C.: Complexity and dynamism from an urban health perspective: a rationale for a system dynamics approach. J. Urban Health. **92**(3), 490–501 (2015)

Vlahov, D., Freudenberg, N., Proietti, F., Ompad, D., Quinn, A., Nandi, V., Galea, S.: Urban as a determinant of health. J. Urban Health. **84**(1), 16–26 (2007)

Vlahov, D., Boufford, J.I., Pearson, C.E., Norris, L.: Urban Health: Global Perspectives, vol. 18. Wiley (2011). Retrieved from https://books.google.com/books?hl=en&lr=&id=Br5oCwAAQBAJ&oi=fnd&pg=PR11&dq=%22urban%22+JI+boufford&ots=UBWSZmVkOW&sig=yHrpLX6NW9SIZ_v1DwlH0O-_qQI

Watts, D.J., Strogatz, S.H.: Collective dynamics of "small-world" networks. Nature. **393**(6684), 440–442 (1998)

World Health Organization: Global Report on Urban Health: Equitable, Healthier Cities for Sustainable Development. UN Habitat (2016). Retrieved from http://www.who.int/kobe_centre/measuring/urban-global-report/ugr_full_report.pdf?ua=1

Zhang, D., Giabbanelli, P.J., Arah, O.A., Zimmerman, F.J.: Impact of different policies on unhealthy dietary behaviors in an urban adult population: an agent-based simulation model. Am. J. Public Health **104**(7), 1217–1222 (2014). http://doi.org/10.2105/AJPH.2014.301934

Zimmerman, F.J.: Habit, custom, and power: a multi-level theory of population health. Soc. Sci. Med. **80**, 47–56 (2013)

Smart Cities IoT: Enablers and Technology Road Map

Larissa R. Suzuki

Abstract The Internet of Things (IoT) is a new paradigm that combines aspects and technologies from ubiquitous and pervasive computing, wireless sensor networks, Internet communication protocols, sensing technologies, communication technologies and embedded devices. Smart cities are advancing towards a pervasive, integrated and intelligent environment, where IoT is used to seamlessly interconnect, interact, control and provide insights about the various silos of fragmented systems within cities. The huge number of interconnected devices as well as the significant amount of data generated by them provides unprecedented opportunities to solve urban challenges. These technologies are merged together with city systems to form an environment where the real and digital worlds meet and are continuously in a synergetic interaction. This intelligent and pervasive environment forms the basis of the interconnected smart cities. In this paper, we elucidate the concept of smart cities, the key features and the driver technologies of IoT and the physical digital integration within city systems. We present smart cities applications enabled by the IoT and research challenges and open issues to be faced for the IoT realisation in smart cities.

1 Introduction

The infrastructure of cities has evolved through many vintages of technology that developed along their own path, often separately. The lack in connecting its component systems, which depend upon the other, often makes city utilities and services operate suboptimally, limiting the creation of new value-added services, increasing transport costs and damaging existing logistics chains and economic

L.R. Suzuki (✉)
Department of Computer Science, University College London, Gower Street, London, WC1E 6BT, UK

Digital City Exchange, Imperial College Business School, South Kensington Campus, Kensington, London, SW7 2AZ, UK
e-mail: Larissa.romualdo.11@ucl.ac.uk

S.T. Rassia, P.M. Pardalos (eds.), *Smart City Networks*, Springer Optimization and Its Applications 125, DOI 10.1007/978-3-319-61313-0_10

models. City planners and decision-makers are therefore increasing their focus on carbon reduction and energy management, as a means of transforming their cities into more efficient and sustainable environment.

Alongside problems in infrastructure, every day nearly 180,000 people move to cities, creating more than 60 million new urban dwellers every year. Cities must plan for population growth and introduce a more sustainable, efficient and liveable model in urban development. Modern digital technologies offer the chance to create a balance between social, environmental and economic opportunities that will be delivered through smart city planning, design and construction. The current proliferation of technologies, open data initiatives and user-generated content is already generating a massive amount of data. The modern city should provide an environment in which information flows rapidly and easily, making of itself a platform for both the dissemination and active consumption of innovation to improve the way it works and peoples' lives. It follows that the systems' operating cities' physical infrastructure needs to become as tightly integrated as they can be, able to draw effectively on a vast supply of cross-domain city data.

The trend for system integration and data collection in city environment is on the rise, as end users seek to realise the benefits of cyber-physical integration through improved control and management of resources. To reduce costs, interconnect activities and integrate systems, cities have increased their reliance on automated machine-to-machine (M2M) interactions (Adams and Jeanrenaud 2008; Nye 1996; Gann et al. 2011). Cities have been infused with devices connected to the Internet, thus creating a network of connected pervasive things, namely, the Internet of Things (IoT).

Within smart cities, the Internet of Things is composed of hardware and software technologies. Hardware components include connected devices (e.g. sensors, metre actuators, smartphones, wearable) and the interconnecting network (e.g. cellular, Wi-Fi, Bluetooth, 5G). Software components include, for instance, vendor's device management systems, data storage platforms and analytic and dashboard applications. The vision of the Internet of Things applied to smart cities is realised when both hardware and software components are interconnected and data is exchanged across different stakeholders, systems and value chains. Such an integrated and intelligent environment, alongside the advancements in technology and advancements in data gathering and analysis, is opening new possibilities for smart cities technology. Converged systems form the basis for the concept of smart cities and realise system-of-system integration and cost-efficient solutions that will provide citizens and businesses with a high quality of life while meeting its ambitious sustainability agenda.

The purpose of this paper is to provide a succinct and systematic review of the main tools and technologies the Internet of Things applications which are transforming cities into a real-time interconnected and smart environment. This paper begins with the discussion of the definition of smart cities. The second section of the paper introduces the concept of the Internet of Things. The third section presents state-of-the-art tools and technologies for the integration of smart cities and the Internet of Things. The fifth section presents applications of the Internet of Things in smart cities. Finally, the last section summarises the emerging research directions, followed by the conclusions.

2 Smart Cities

The world's population is growing at a rapid pace. Whereas in the early twentieth century only 13% of the world's population lived in cities, this ratio amplified to 29% in 1950 and to 50% in 2009 and is expected to reach 70% by 2050. This unprecedented speed in urbanisation in conjunction with the current economic crisis has become an overwhelming issue for city governance and politics cities. Cities and towns are suffering to provide basic urban services, and the increasingly consumption by humans has amplified scarcity of environment and natural resources in the earth. Among the various challenges that city planners, businesses and governments will have to be concerned about are the provision of a more sustainable, secure and affordable energy, better infrastructure to manage energy supply and peak demand and meet CO2 reduction targets, more secure and high-quality water, more integrated public transport as well as the optimisation of the existing assets. Five decades after the emergence of the digital computer, modern digital technologies offer new opportunities to manage urban transformations (Hall and Pfeiffer 2000) and foster the era of sustainability (Adams and Jeanrenaud 2008).

Alongside with environmental changes, inadequate, deteriorating, fragmented and ageing infrastructure severely affects the health of cities (Nye 1996; Gann et al. 2011; van den Besselaar and Beckers 2005). Cities are composed of many fragmented information systems and applications which developed along their own path, often separately. The lack in connecting its component systems makes city utilities and services operate suboptimally, limiting the creation of new value-added services, increasing transport costs and damaging existing logistics chains and economic models. Problems in cities' infrastructure have often been accompanied by the aggravation of many challenges associated with urban living in terms of law and order, health, waste disposal, housing, utilities, education, transportation and delivery of basic public services (Suzuki 2015a).

The aforementioned challenges and complications have motivated policymakers to seek balances between industrialisation, economic development, urban growth, geographic sprawl and environmental necessities to create sustainable cities. As a main goal, cities are looking for intelligent solutions that will improve the efficiency of urban services and provide a high quality of life without excessive operating costs.

Digital technologies offer a new wave of opportunities to mitigate some of these impacts and create a balance between social, environmental and economic opportunities that will be delivered through smart city planning, design and construction. As cities seek to reduce costs and interconnect their activities, urban systems have become tightly integrated and increasingly relying on real-time cross-domain data.

Real-time city data have enabled the creation of several sustainable initiatives and real-time services in cities, such as smart electricity grids (Klein and Kaefer 2008) and public space monitoring (Filipponi et al. 2010), among others.

The vision of using digital technologies to optimising existent cities' infrastructure instead of developing new cities from scratch has originated the science of smart cities. City policymakers and organisations are increasingly inclined in creating smarter cities to orchestrate cities' activities and integrate their component systems. They expect that smart cities will fuel sustainable economic development and boost economic growth. This new science of smart cities aims to demystify how city systems (e.g. energy, water, transport, environment, waste and recycling, healthcare) can work in an orchestrated manner through data exchange to make a city a better place to live, where business can prosper, citizens are happy and healthy and there is a sustainable economy (Suzuki 2015a). Smart cities represent a very important new concept, and it is no surprise that it has received so much attention from researchers, private organisations, government and investors.

2.1 Smart Cities Drivers and Enablers

One of the main challenges found in the management of urban environment is how to promptly react to unpredicted emergency situations. Often, immediately real-time reaction can be limited because of the fragmented scenario found in city systems, lack in connecting sensors and actuators and businesses' processes and services which must change to allow immediate response to the emergency. As city systems, processes and services have not been properly interconnected, immediate reaction in response to emergencies and the efficient management of urban resources is limited.

The current advancements in modern digital technologies and the decreasing costs on computation power will significantly change the way systems operate cities' infrastructures. Once data is collected, the city physical infrastructure is infused with information, and data can be integrated and exchanged across multiple processes, entities and systems and used to recover rapidly from disasters and help cities to meet their sustainability targets. Digital infrastructures and users may access city data to make intelligent decisions, improve service levels, better manage infrastructure and assets as well as meet sustainability targets.

The high availability of city data and integration of city systems will be the differentiation factor for unlocking value from data that will support intelligent decisions in the urban environment. The integration of city systems at the system-of-system level has been demonstrated to be able to create drivers for infrastructure innovation and improve the control of resources (Gann et al. 2011). Consequently, the city infrastructure and systems that were previously invisible can then be discovered and improved through integrated city data.

In the context of this paper, smart cities solutions have been driven and enabled by both technology and social engagement:

- *The social networks' and citizens' data*: Citizens will be responsible for a great part of the huge volume of data expected to arise in cities. This data will mostly come from weblogs, social media, mobile devices, automobiles, smart cards, smart homes (home area network, HAN) and others. The heterogeneity of devices connected to HANs is expected to increase rapidly in the next couple of years as a consequence of the global interest in creating more "sustainable" energy devices such as smart refrigerators, thermostats and smart metres (Klein and Kaefer 2008).
- *Cloud computing scalability*: Elasticity, pay-as-you-go solutions for computing cycles requested on demand, low upfront investment, low time to market and transfer of risks are some of the major enabling features that make cloud computing a ubiquitous paradigm for deploying novel business models which were not economically feasible in a traditional enterprise infrastructure setting. This has seen a proliferation in the number of applications which leverage various cloud platforms, resulting in a tremendous increase in the scale of the data generated as well as consumed by such applications (Zhang et al. 2010).
- *Government open data*: Open data initiatives have released a vast amount of data (transport, water, environment, geospatial) for public access (Shadbolt et al. 2012). The key fundamentals of Web 2.0 technologies include the idea of reusable, open data and open interfaces (APIs), so that data from one service can be combined with data from another to create interesting data combination and integration. Government open data is becoming very popular, and governments are looking for ways to increase transparency and accountability, to improve efficiency and to contribute valuable information to the constituents (Janssen et al. 2012).

Realising smart cities will necessarily create demand for an ample integration with numerous external resources, such as data storages, services and algorithms, which can be found within organisation units, other organisations or on the Internet. The main objective when interconnecting devices (e.g. sensors) and collecting and processing their data is to create situation awareness and enable applications, machines and human users to better understand their surrounding environments. Therefore, this understanding can potentially enable services and applications to make intelligent decisions and to respond to the dynamics of their environments. The diversity, instability and ubiquity in the sensory data make the task of processing, integrating and interpreting the real-world data a challenging task.

Another issue is the massive data pool being created within cities which will shortly create new data ingest, aggregation and analytic challenges that will exceed the limits of vertically scalable tools requiring distributed storage solutions, parallel processing tools and very-high-speed and capacity networks. This voluminous information environment has initiated a new era of big data in smart cities.

To interconnect their physical infrastructure and cross-domain data, cities will need to tightly integrate their systems and increase their reliance on automated machine-to-machine (M2M) interaction (Karvonen 2001). This high level of interconnectivity will create ever more complex systems of systems that have the potential to efficiently and effectively manage resource in cities.

2.2 Physical-Digital Integration in Smart Cities

Smart cities are composed of many fragmented information systems and applications which provide services for citizens, businesses and public departments (Suzuki 2015a). Among these fragmented systems included geographic information systems, intelligent traffic systems, healthcare, buildings and constructions, security, transportation, water and waste supply, environmental monitoring systems, energy and electricity. In addition, cities are also equipped with heterogeneous devices and have different stakeholders. Figure 1 illustrates the city as a system of systems.

Fig. 1 Cities as a System of Systems (Suzuki 2015a)

Cities are composed of several core systems, each on different networks, infrastructure and environments. Basically, they can be described as government/city services, environment and urban planning, education, business, water and energy, transport and public security. Although city systems provide control over critical resources, they are very expensive and hard to customise, maintain and extend. This is due mainly to the closed and proprietary nature of such systems, which are often developed/operated using different languages and data formats, varying in terms of architectural design and communication protocols (Gann et al. 2011; Suzuki 2015a), and are geographically distributed and exhibit emergent behaviour. This situation results in city operators to expend considerable effort to manage multiple independent systems. Consequently, problems have been reported with achieving full utilisation of such systems, and many cities struggle to share information across different silos.

In order to reduce costs associated with proprietary systems solutions and integration burdens, city systems have increased their reliance on automated M2M interactions and tied to a pool of semi- and nonstructured real-time data which is catalysed by millions of electronic networked devices responsible to manage and operate the physical infrastructure of cities (e.g. sensors, smart metres, cameras and actuators). This network of connected devices is known as the Internet of Things (IoT) (Atzori et al. 2010). The IoT is one of the key low-cost technologies that has the power to manage, plan, interconnect and operate assets without the need to retrofit existing structures.

The Internet of Things refers to digital networks capable of interconnecting infrastructure, data and people through various digital devices. It is estimated that 50 billion such devices will be connected by 2020. Due to IoT's interconnecting power and generation of voluminous amount of key emerging data, it will soon become a technology providing major transformation in public, private and community services and interactions.

3 The Internet of Things

The Internet of Things (IoT) emerged at MIT around the year 2000 (Ashton 2009), and it is defined by the International Telecommunication Union (ITU) as "a new dimension has been added to the world of information and communication technologies (ICTs): from anytime, anyplace connectivity for anyone, we will now have connectivity for anything. Connections will multiply and create an entirely new dynamic network of networks – an Internet of Things". It is expected that by 2050, over 50 billion devices will be connected to the Internet, representing 6.58 devices per human (Fig. 2).

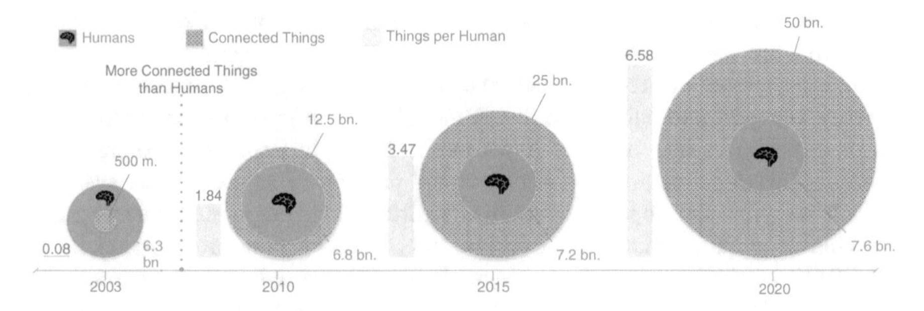

Fig. 2 The dawn of connected machines (Suzuki 2015b)

The IoT has created an entirely new dynamic network of networks which has revolutionised the world with the idea of the connectivity of anything, from anyone, at anytime and anywhere. In the IoT paradigm, things can be sensors, actuators, lifts, lights, mobile phones, wearables and so forth. These things have identities, physical attributes and virtual personalities, use intelligent interfaces and can interact and collaborate within their environment using unique addressing schemes in order to achieve common goals (Atzori et al. 2010). Besides smart objects, machine-to-machine (M2M) communication is another fundamental component of the IoT. M2M is a low-power and low-cost association of computing technologies and communication via smart and pervasive objects which interact without human intervention, with the information systems (Lawton 2004).

Hence, the concept of IoT aims at making cities even more interconnected, immersive and pervasive. By allowing easy access and interaction with a wide variety of devices (e.g. sensors, actuators, appliances, vehicles, displays, appliances, mobile phones), the IoT will foster the development of new integrated services to citizens, businesses and public administrations. This paradigm indeed finds application in many different domains, such as smart homes, smart buildings, healthcare, elderly and visually impaired assistance, smart grids, automotive, traffic management and many others that will be discussed in Sect. 4.

An increasing number of IoT applications have emerged recently, and according to Atzori et al. (2010), they can be categorised as the following, and the IoT is the result of the convergence of these different visions:

- *Things oriented*, where the focus is on identifying and integrating objects
- *Internet oriented*, where the emphasis is on establishing an efficient connection between devices and exploring the applications of IP protocol
- *Semantic oriented*, which is concerned with describing objects and managing the huge amount of data provided by the increasing number of IoT objects

However, such a heterogeneous field of application makes the identification of solutions capable of satisfying the requirements of all possible application scenarios a formidable challenge. This difficulty has led to the proliferation of different and, sometimes, incompatible proposals for the practical realisation of IoT systems.

Therefore, from a system perspective, the realisation of an IoT network, together with the required backend network services and devices, still lacks an established best practice because of its novelty and complexity. In addition to the technical difficulties, the adoption of the IoT paradigm is also hindered by the lack of a clear and widely accepted business model that can attract investments to promote the deployment of these technologies (Delaney et al. 2009).

3.1 The Horizontal Integration of the IoT

In the last decades, city's legacy systems have been primarily designed for specific purposes with no or limited flexibility; they often do not share any features for managing services and network, resulting in unnecessary redundancy and increase of costs. This entirely vertical approach should be overtaken by a more flexible and horizontal approach, where a common operational platform will manage the network and the services and will abstract across a diverse range of data sources to enable anytime and anywhere access to applications. In the context of smart cities, there is an increasing demand for service adjustment (e.g. control of resources) to accommodate application and service platforms which can capture, communicate, store, access and share data from the IoT.

Figure 3 illustrates applications working in isolation and the vision of shared infrastructure and platforms that can integrate and orchestrate the overall architecture. In this figure, there are three different layers in which integration takes place. Firstly, (1) at the lowest layer where the physical devices reside and where data is collected, (2) the next layer up where data is transmitted over the network and (3) the top layer where applications manage and use data gathered from the physical environment. Some examples of technologies and protocols used in each layer are also shown in Fig. 3. A more detailed view on such technologies and protocols is given in the next section.

Most IoT solutions rely on wireless sensor networks (WSNs) which are a powerful technology for gathering and processing data in a large variety of domains, from environmental monitoring (Al-Turjman et al. 2013) to smart grids (Lu et al. 2010). Traditional WSNs consist of a high number of static and resource-constrained sensor nodes deployed in an area to sense a certain phenomenon, e.g. temperature and humidity.

The instrumentation of smart cities with IoT technologies enables real-time data to be collected and infused in the physical infrastructure to enable efficient monitoring of resources and prompt reaction to unpredicted situations. For example, energy, water, oil or gas metres can measure energy consumption in much more detail than conventional metres and offer two-way (near or real-time) information transmission between the customer and the authorised parties (e.g. utility providers, service operators, etc.). The infrastructure and the systems that were previously invisible can then be discovered and improved through digital data.

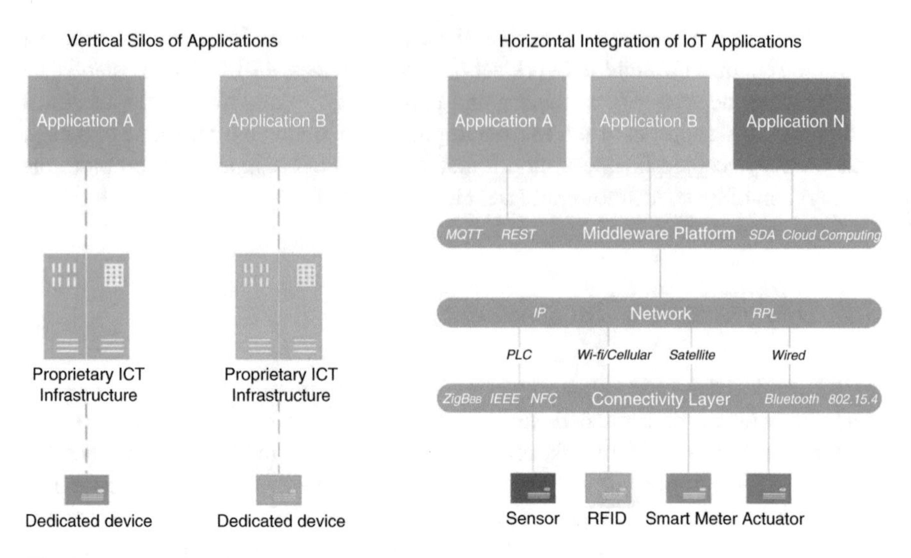

Fig. 3 Vertical silos of applications vs horizontal integration of IoT applications

4 The Internet of Things Technologies for Smart Cities

The realisation of smart cities involves creating, processing and communicating information, and therefore data has a pervasive influence in its design. Smart technologies and integrated systems offer opportunities for systems to communicate with other systems and to adapt to conditions imposed by them and many others (Myers 1996; Trankler and Kanoun 2001; Wang and Xie 2002). The efficient management and integration of city component systems and IoT requires technology capable of supporting a wide range of devices and systems (e.g. sensors, appliances, actuators, smart grids, etc.) together with the interests of smart cities planners and decision-makers.

Smart cities system integration takes place at physical, network and application levels. This integration involves interconnecting systems both physically and functionally in a common architecture to share and exchange data (e.g. Myers 1996; So and Tse 2001). The physical dimension refers to cabling, spaces, power and the use of open protocols. Functional dimension refers to the provision of capabilities that cannot be achieved by any system in isolation. Integrated systems aim at having a single database in order to reduce cost and support for synchronising separate databases, which happens in the case of systems that are interfaced rather than integrated (Suzuki 2015a).

Alongside the several fragmented systems embedded in city settings, many M2M solutions (e.g. sensors and actuators) and WSNs have been increasingly embedded in city environment with the purpose to enable such complex environment be managed holistically. Although the Internet enables smart objects, low-powered devices and WSNs to grow beyond narrower domains/fields of deployments, it imposes limitations as smart objects suffer from power/data loss constraints and scalability limiters. Hence, realising IoT systems that are low cost, auto-organising, easily accessible, efficient and energy aware requires adopting efficient standards.

IoT deployments in smart cities are complicated due to different communication standards, from wireless protocols such as ZigBee and Bluetooth to next-generation protocol standards (e.g. 802.15.4e, 6LoWPAN) which attempt to unify the wireless sensor networks and the established Internet. For instance, IoT applications for smart cities require network technologies that support high data rate resulting from the aggregation of an extremely high number of smaller data flows. For these reasons, smart cities applications require the adoption of different technologies at the network layer, and they are classified into:

- Non-constrained technologies such as LAN, WAN, fibre optic and Wi-Fi which are not suitable for resource-constrained devices at the edge of the IoT given its high complexity and high energy consumption
- Constrained technologies such as IEEE 802.15.4, Bluetooth, NFC and RFID (Ko et al. 2011) which exhibit low energy consumption and transmission rates

Many groups have been created to provide protocols in support of the IoT including efforts led by the World Wide Web Consortium (W3C), *Institute of Electrical and Electronics Engineers* (IEEE), International Organization for Standardization (ISO), *Internet Engineering Task Force* (IETF) and *European Telecommunications Standards Institute* (ETSI). However, in this paper, we focus on constrained technologies and IoT technologies in the context of sensors, and therefore other technologies such as RFID and NFC are not discussed.

4.1 Standard IP IoT Protocol Suite for Smart Cities

The Open Systems Interconnection model (Day and Zimmerman 1983) depicts the flow of information within an open communication networks in seven distinct layers. Figure 4 illustrates the OSI model layers, the Internet protocol suite and the IP IoT protocol suite for smart cities. In the next sections, we present the main technologies and protocols used in integrating smart cities systems and the Internet of Things.

4.1.1 Physical Layer

The first step towards IoT is the connection with the physical layer from which we can collect information from or about objects (e.g. identity, state, energy level). In the IoT protocol suite (Fig. 3), IEEE 802.15.4 standard (Tiberi et al. 2010) is a physical layer standard that supports full function devices (FFD) and reduced function devices (RFD), types of network nodes. FFD can serve as a normal object in a network or perform more complete roles such as being a personal area network

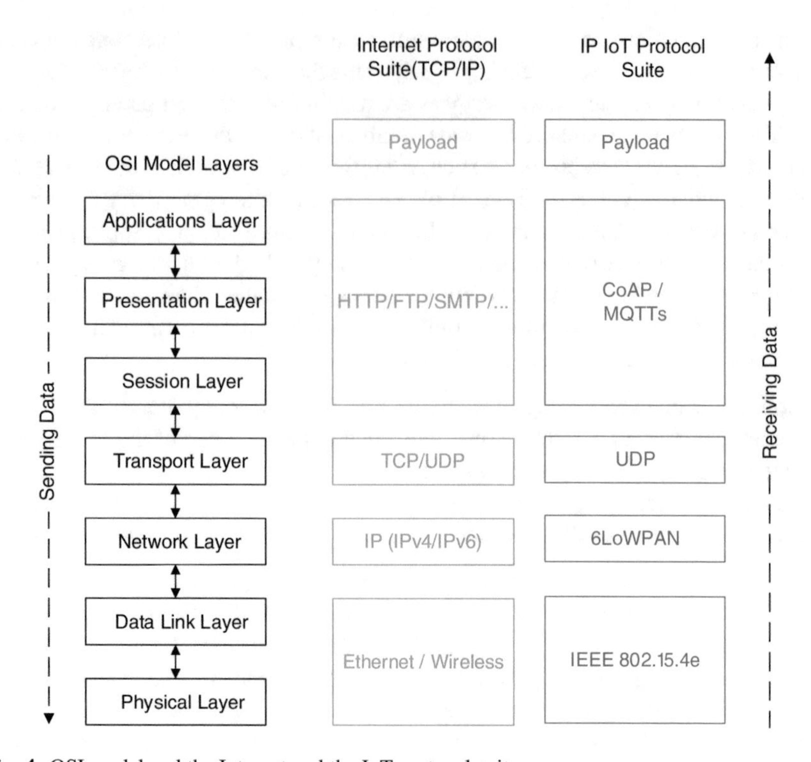

Fig. 4 OSI model and the Internet and the IoT protocol suite

(PAN) coordinator. RFD are resource-constrained embedded devices in low-power and lossy networks (LLNs) with fixed, portable and mobile objects with none or limited battery consumption requirements.

Among the various characteristics of IEEE 802.15.4 are it has low data transmission, high message throughput, low frame overhead, low cost and efficient power management mechanisms and is the foundation of several well-known protocols like ZigBee, 6LoWPAN and others. Due to such specifications, it is the typical physical layer protocol used by the IoT, M2M and WSNs. Although it provides a reliable communication, operates on different platforms and has high level of security (including encryption and authentication capabilities), it does not provide QoS guarantees.

4.1.2 Network Layer

The Internet of Things is expected to connect billions of devices to the Internet by 2020. In theory, each IoT device will need to be uniquely addressable. IPv4's capability to assign IP addresses globally has exhausted, and the Internet Protocol version 6 (IPv6) (Hui and Thubert 2011; Montenegro et al. 2007) has been introduced as a suitable alternative due to its 128-bit address field capability, thus making it possible to assign a unique IPv6 address to devices in the IoT networks.

However, at the IoT network layer, the IPv6 is not suitable for use as it introduces a very big overhead when communicating with low-power radios and low data rate links (resource constrained) such as IEEE 802.15.4 (Montenegro et al. 2010). To overcome this issue, the 6LoWPAN (IPv6 over low-power wireless personal area networks) header compression format was created to enable efficient transmission of IPv6 packets over LLNs (Zach and Carsten 2009; Mulligan 2007). This open standard is supported by IETF (Internet Engineering Task Force) and enables devices to be directly connected to the web. This protocol introduces an adaptation layer between the IP stack's link and IEEE 802.15.4 specifications and has limited packet size, different address lengths and low bandwidth (Ko et al. 2011). Data rates are available at 20 kb/s, 40 kb/s and 250 kb/s.

4.1.3 Transport Layer

Most of the Internet traffic is carried at the application layer by HTTP over the Transmission Control Protocol (TCP). TCP which is used for most of our human interactions is not useful for IoT applications as the message overhead introduced in messages yields poor performance for small data flows in embedded systems.

Above the network layer, the UDP transport protocol (Nikander et al. 2010) is used between the 6LoWPAN (network layer) and the application layer. Unlike TCP, the UDP protocol uses a simple transmission model avoiding a big overhead. However, UDP does not have a built-in error recovery mechanism and lacks reliability, and as such, these issues must be addressed in other layers to ensure correct delivery of packets.

4.1.4 Application Layer

1. The *Constrained Application Protocol (CoAP)* was developed by the IETF-Constrained RESTful Environments (CoRE) working group for IoT applications (Bormann et al. 2012). CoAP (Shelby et al. 2012) is bound to UDP and has been designed with IoT requirements in mind, adding reliability mechanisms and targeting at resource-constrained devices and at the REST paradigm (Fielding 2000) for objects addressing on top of HTTP functionalities. CoAP modifies some HTTP functionalities to ensure low-power consumption and operation in lossy and noisy environment. CoAP identifies resources using a universal resource identifier (URI) and defines a web transfer protocol based on REST allowing the resource to be affected using similar methods such as GET, PUT, POST and DELETE. Furthermore, CoAP's reliability of communications is guaranteed by QoS mechanisms based on acknowledgements. It utilises four types of messages: confirmable, non-confirmable, reset and acknowledgement.

2. The *Message Queuing Telemetry Transport (MQTT)* protocol (MQ Telemetry Transport n.d.) is a viable solution for time-critical applications which require reliability. MQTT is a topic-based publish/subscribe communication protocol that is designed to be open, simple, lightweight, easy to implement and efficient in terms of processor, memory and network resources. MQTT is built on top of the TCP protocol and provides routing for small, low-cost, low-power constrained devices in unreliable or low bandwidth networks. MQTT delivers messages through three levels of QoS. The MQTT-SN (Hunkeler et al. 2008) (formerly known as MQTT-S) has been designed keeping low-end and battery-constrained sensors/actuators in mind by defining a UDP mapping of MQTT and a broker support for indexing topic names. MQTT simply consists of three components, subscriber, publisher and broker. In this environment, a device subscribes for specific topics; it has interest whose data is provided by the broker when the publisher publishes data on that topic. The publisher, on the other hand, transmits the data to the interested and authorised subscribers via the broker. The protocol possesses QoS mechanisms for message delivery assurance.

The high scalability of the IoT requires mechanisms for resource management which enables services to be registered and discovered in an efficient, self-configured and dynamic way. The prevailing protocols for service discovery and registration in IoT are versions of multicast DNS (mDNS) (Cheshire and Krochmal 2013) and DNS service discovery (DNS-SD) (Krochmal and Cheshire 2013) adapted to resource-constrained environments (Jara et al. 2012).

4.2 Security Elements of IoT

Security issues are central in IoT as they may occur at various levels. To secure the entire IoT system, confidentiality, integrity, authentication, authorisation, non-repudiation, availability and privacy must be guaranteed (Heer et al. 2011; Cirani et al. 2013). This is extremely challenging since to provide good performance of the systems, the packets transmitted are fragmented which may simplify network attacks. Secure identification of communicating devices, secure data storage, authentication and authorisation can be provided by SIM cards and M2M SIM cards. These technologies use PIN, PUK and public key infrastructure (PKI) and require installing and configuring credentials, keys and certificates on the devices (Jennings 2012).

Security issues affecting data communications are found in the network, transport and application layers. The most used secure Internet transport protocols for the IoT is Datagram Transport Layer Security (DTLS) which run over UDP (Nikander et al. 2010) (Table 1). Cryptographic algorithms are also used to ensure the integrity of data travelling through the network (Eisenbarth and Kumar 2007; Standaert et al. 2006; Bogdanov et al. 2000).

Table 1 IoT communication protocol

Application protocol	Transport	Security	QoS
REST/HTTP	TCP	SSL/TLS	No
CoAP	UDP	DTSL	Yes
MQTT	TCP	SSL	Yes
MQTT-SN	TCP (UDP mapping)	SSL	Yes

4.3 Data Interoperability and Semantics in IoT for Smart Cities

To manage smart cities holistically and model the implications of connections across their component systems, it is becoming essential to derive knowledge from across different data sources and domains. New IoT technologies deployed at a city scale will yield ever-increasing amounts of data from existing and new sources which analytics can mine for intelligence and provide real-time situational awareness. Two key aspects in producing insights are how to bring high-quality voluminous heterogeneous data sources together and how to reuse the data across different value chains and stakeholders.

However, there is often little interaction between the traditional islands of city data (e.g. legacy systems, SCADA) and emerging systems, and even where interconnection can be achieved, the potential range and volume of data can quickly become overwhelming without a clear and focused methodology for data management. Interoperability of emerging technologies and data management tools is therefore a key concern.

Data integration needs for smart cities data can be summarised in a series of methodological steps. Once data is collected from systems and devices, data should undergo many pre-processing steps for the verification of the completeness, provenance and quality. Provenance can be used to give a historical account of when and how data has been produced, its ownership and all the necessary information to reproduce the results. One of the most difficult tasks involves how to automatically generate the right metadata, how to describe the data and how it is recorded and measured. The data must be semantically annotated and modelled in a standard and established format to enable data to be both human and machine readable and understandable. Sensitive information regarding users, devices and systems needs to be protected at this stage to assure data privacy. Once the data is modelled, it can be linked to other internal or external data repositories to enable data discovery and reuse. The ability to reuse data by not having to collect the same data again alleviates unnecessary duplication of data and associated costs (Suzuki 2015a).

Moving beyond the integration of data described, a further challenge exists in confirming how best to manage the overall process in such way that the data is easily stored, processed and retrieved across different systems and value chains and infused back into smart building systems to generate informed and automated decisions. This requirement is supported by semantic web technologies such as the Resource Description Framework (RDF), the Web Ontology Language (OWL) and the linked data (Heath and Bizer 2011). The semantic web enables automatic interoperation between data provided via the web, which can be expressed in a way that facilitates its understanding by computing machines. Berners-Lee and Fischetti (Berners-Lee and Fischetti 1999) describe the semantic web as "an extension of the current Web where information is given a well-defined meaning".

A key effort in this domain is the linked data model (Heath and Bizer 2011) to overcome the barriers of data interoperability and the interconnection of disparate and fragmented silos of information. Linked data is expressed in open and nonproprietary formats, with modular approach to data instances so that datasets can be combined (mashed-up) with any other pieces of linked data. Web Ontology Language (OWL) is based on RDF and refers to a semantic web technique used to model the meaning of data (the context) and can be defined as formal and explicit specifications of a domain which are shared among many stakeholders and can be used for inference and reasoning. Linked data can be queried using a language and protocol called SPARQL which is the standard RDF query language (Heath and Bizer 2011). For its capabilities to publish, connect, integrate and share data, linked data constitutes a powerful tool to model smart city big data as it provides means to create a connected and discoverable data environment. As linked data works at the application layer and not at the physical infrastructure level (discrete hardware standards/software protocols), system architectures can be designed independently and later linked at the edges, building in interoperability incrementally where it is most needed and most cost-effective. This approach allows for scalability even when the terms and definitions of the datasets used change over time, while also enabling data exchange and reuse across different systems and by a wide range of stakeholders.

5 IoT Applications in Smart Cities

Several efforts from industry and academia have been placed on Iot technologies applied to smart cities. Among applications enabled by the IoT, we highlight smart homes, smart buildings, smart healthcare, intelligent transport systems, smart grids and linked data platforms for urban data management.

Examples of such solutions are architecture for video sensing in cities (Wen et al. 2010), management and cooperation of heterogeneous sensors for public space monitoring (Filipponi et al. 2010), efficient use of electricity within smart infrastructure and data centres (Klein and Kaefer 2008), real-time architecture for disaster management based on information collected from various entities (e.g. crowdsourcing, homes, vehicles) (Asimakopoulou and Bessis 2011), physical and technical countermeasures against the security threats (Jung et al. 2009) and pipeline monitoring via the deployment of sensors around the pipeline for continuous monitoring (Metje et al. 2011).

Other solutions are more focused on sensor networks, the Internet of Things and the Internet of Services. Smart homes powered by the IoT enable everyday household electronics and smart appliances to communicate with one another either locally or via the Internet (Elkhodr et al. 2015). Ultimately, smart homes have been designed to enhance the personal lifestyle of occupiers by making their homes and personal spaces easier and more convenient to monitor, operate and optimise the usage of resources (e.g. air conditioner, heating, energy, water, etc.). Research in this area includes the integration of home and IoT devices with data aggregation, reasoning and context awareness capabilities to monitor domestic conditions regularly (Kelly et al. 2013). Another stream of research, namely, Ambient Assisted Living (Skouby et al. 2014), focuses on monitoring the well-being of residents. Ghayvat et al. (2015) propose to track the daily routine of elderly home occupants and monitor their activities for health purposes, while Zhang et al. (2013) proposed a solution to maintain users' awareness of healthy postures during their daily activities. Nasir et al. (Nasir et al. 2015) discuss a cognitive network for monitoring and predicting the water usage in smart homes using RFID sensors. Their solution enables the prediction of optimal usage of the water resources in smart homes. Kamilaris et al. (2011) propose a solution for a web-based energy-aware smart home framework that enables smart appliances to be connected on the web and the overarching smart home to be connected to smart grids.

Research on smart buildings have been devoted on the advanced development of system integration approaches, occupancy (Erickson et al. 2009) and building subsystem services (Brennan et al. 2009), such as HVAC systems (Lüthi et al. 2001), and managing IT energy consumption (Agarwal et al. 2009) within buildings, and IoT applications. Emergency management solutions have received considerable attention in smart building research. Such solutions are capable of assisting building operators in preparing for and coping with natural or man-made disasters such as chemical leaks, floods, fire, earthquakes and electrical outages. Dedicated sensors and intelligent cameras, as well as GPS and wireless technologies providing real-time localisation (Deak et al. 2012) and tracking (D'Souza et al. 2013), can be used to form a complete map of the event, to forecast its trends (e.g. direction and/or speed of fire spread, major risk areas) and thus to establish a dynamic emergency plan to coordinate the rescue operations. Several approaches focusing on using a user's preference data to manage buildings consumption have also been proposed. For instance, Singhvi et al. (2005) propose a solution to optimise the trade-off between meeting user comfort and reduction in operation cost by reducing energy usage. Chen et al. (2009) propose a smart building architecture that keeps track of workers' real-time location in an office and retrieve their personal preferences of lighting, cooling and heating.

In the area of ontologies and semantics of IoT technologies (Compton et al. 2009; Eid et al. 2007), different ontologies and semantic models are presented for sensor data representation. To understand how smart cities may benefit from IoT technologies, Hernandez-Munoz et al. (2011) presented an extension of their framework, called ubiquitous sensor network (USN), in which services could be developed at minimal cost. This solution enables the integration of heterogeneous and geographically dispersed sensors in a centralised technological base. The authors highlight the need to develop advanced services for data filtering and aggregation. Andreini et al. (2011) proposed a service-oriented architecture for M2M interactions, in which objects can publish their services to be accessed by mobile phones. This architecture uses techniques to improve the scalability of the system and rapid recovery of services for smart cities applications. Live Singapore project (Ratti and Townsend 2011) is supported by modern pervasive technologies to monitor urban environments and assist on decision-making process. Milton et al. (2012) propose sensing as a service approach for smart cities where sensors and actuators not only can be discovered and aggregated but also dynamically provide sensing as a service, applying the cloud provisioning model. Attwood et al. (2011) proposed a framework to support real-time analysis, visualisation and modification of a failing system within a smart city. In case of failure in any sensor of the network, another overlay network would be created from sensor nodes still in operation, allowing data preservation and distribution. Linked data applications have been recently applied to the IoT (Le Phuoc and Hauswirth 2009; Wang et al. 2013), in which software middleware is designed to integrate and share urban data streams from IoT using linked data and common vocabularies. However, the authors report on the high cost that converting data to RDF introduces in the platform and that new approaches are needed to semantically annotate and convert files into RDF format as well as tracking data quality and a seamless data integration.

6 Open Challenges and Road Map

Internet of Things applications can perform the same function as other systems distributed in the city environment. For instance, IoT and existing systems can both control traffic lights and energy systems, and they can both be operated remotely. When it comes to what IoT can perform beyond what existing system can, then the true power of the Internet of Things in smart cities and its economic benefits can be comprehended. For instance, often the fragmented systems operating and managing the physical infrastructure of cities are very costly to execute and maintain. Such systems are often too complex to be operated by a nonexpert, requiring engineers to operate them and therefore incurring engineering costs. Furthermore, these systems also require human input and frequently rely on manual data processing, integration and parameterisation of variables, thus limiting buildings in achieving an optimal operation level. The IoT solves these problems as its capability goes beyond simple preprogrammed settings and creates intelligent and context-aware applications that can adapt to conditions and bring city services to the optimal operation level.

The IoT can collaborate with existing systems distributed across cities per inputs received from the environment. The context-awareness concept allows city systems to be truly cost conscious as it can automatically change traffic lights because of accidents or jams, trigger the evacuation of environments due to alarms of security threats and so forth. Hence, IoT applications can assist cities to reduce costs associated with maintenance, integration and processing of data in fragmented systems and create a better environment through integrated services and data.

Supported by an innovative, vibrant and developed community, IoT is expanding the connectivity and scale of machines and sensors throughout the world. Taking advantage of the resulting economies of scale, flexibility and openness, smart cities which make use of IoT technologies are well positioned to reduce solution costs by an order of magnitude compared to today's proprietary and closed systems used in city vertical systems (e.g. transport, energy, water, waste and recycling). Solutions can be tailored to meet the specific requirements of different types of city systems using open standards and communication protocols. As greater reliance is placed on the Internet of Things as the global infrastructure for generation and gathering of information, it will be essential to ensure that international quality and integrity standards are deployed and further developed, as necessary to ensure that the data can be trusted and traced to its original authentic sources. In this context, a close collaboration among different standardised institutions and other worldwide interest groups and alliances is mandatory.

Despite the high potential of such applications, security, trust and interoperability problem remain an open issues. IoT applications in smart city raise many challenges in security and privacy since users implicitly expect their data to be secure and privacy-preserved. With regard to privacy, there is great public fear regarding the inappropriate use of personal data, especially when data is being shared with third parties. It is of utmost importance to address privacy issues to get the most of the big data promise. Some examples of sensitive data found in cities are transport historical data which contains user personal details and transport usage patterns, surveillance data (videos and images), smart metering data, payment of parking and bicycle hiring. There are several privacy and security concerns surrounding smart metering data, and most of these concerns arise from the main functions of the metre itself (McDaniel and McLaughlin 2009). An additional challenge to privacy in ubiquitous environments is introduced by context awareness. The ability of the environment to detect, reason about and act in accordance to the users' context has great potential for invasion of their privacy. The disclosure of context information might allow harmful inferences such as where users are and where they are not (Terzis et al. 2005).

Another vulnerability issue is the level of access granted to data. As there are many different stakeholders within a city, this issue must to be addressed to grant access to data for only those stakeholders to whom the data is of relevance. Furthermore, when data is aggregated, different access conditions may apply to the aggregated data than that found for the individual records. In addition, by centralising data in one place, it becomes a valuable target for attackers which can potentially leave huge swathes of information exposed. It could potentially undermine trust in the organisation and damage its reputation. This makes it essential that IoT data is properly controlled and protected.

There are calls for novel reliable transport protocols to cope with congestion issues that may arise due to the scale of the network, while considering that

M2M communications are short-lived and involve the transmission of few bytes per message. Furthermore, the development of artificial intelligence applications powered by IoT in smart cities remains an open issue. The Internet of Things offers limited processing ability, limited power resources and limited bandwidth resources (Thomas). Advancements in 5G wireless communication technology (Simsek et al. 2016) and network communications will provide additional enabling connectivity and increase in data rate that will enable the Internet of Things to be formed by an extensive variety of powerful end user, network edge and access devices such as smartphones, smart home appliances, connected vehicles, wearable devices, cellular base stations and edge routers, among others. Research on the Tactile Internet (Maier et al. 2016) is advancing technologies that capture and reproduce various stimuli (e.g. sight, hearing, touch, smell and taste) from the outside world and let humans as well as machines perceive and react to the combined stimuli in various ways.

These new technologies have a huge potential to be part of the Internet of Things, and data collected from such technologies provides unprecedented opportunities to solve social challenges in smart cities. However, the data volume is expected to exceed the design specifications and limits of existing networking systems and today's cloud and host computing models. For instance, for health monitoring, emergency response and time-sensitive control functions for cyber-physical systems and other latency-sensitive applications, the delay caused by transferring data to the cloud for storage and processing and back to the application may require prohibitively high network bandwidth impairing scalability or may sometimes be prohibited due to regulations and data privacy concerns.

To address the efficient use of underlying resources, edge computing (Shi and Dustdar 2016) has been proposed to use computing resources near IoT sensors for local storage and preliminary data processing. Although this approach has the potential to decrease network bottlenecks and accelerate analysis, edge devices, however, have not demonstrated to cope with multiple IoT applications competing for their limited resources. This issue may result in increased processing latency caused by resource contention and, consequently, undermining the potential of time-sensitive applications in smart cities. Platforms and software that facilitate the development of more efficient, flexible, reliable, secure and interconnected Internet of Things applications for smart cities remain an open challenge.

As IoT is an emerging paradigm, it is highly dynamic and continuously evolves. Although we have overviewed the main IoT research area as well as related IoT challenges, many other research issues can be identified. The main issues to be addressed are IoT standardisation, protection over personal data, semantic ontologies' standards, communication protocol standards and provenance, as well as standards for communication within and outside cloud computing technologies.

References

Adams, W., Jeanrenaud, S.: Transition to Sustainability: Towards a Humane and Diverse World. IUCN, Gland (2008)

Agarwal Y., Hodges S., Chandra R., Scott J., Bahl P., Gupta R.: Somniloquy: Augmenting Network Interfaces to Reduce PC Energy Usage. In Proc. of USENIX Symposium on Networked Systems Design and Implementation (NSDI '09) (2009)

Al-Turjman, F.M., Hassanein, H.S., Ibnkahla, M.A.: Efficient deployment of wireless sensor networks targeting environment monitoring applications. Comput. Commun. **36**(2), 135–148 (2013)

Andreini, F., Crisciani, F., Cicconetti, C., Mambrini, R.: A scalable architecture for geo-localized service access in smart cities. In Future Network and Mobile Summit, 1–8 (2011)

Ashton, K.: That "The Internet of Things" thing. RFID J Int Telecommun Union, 'ITU Internet Report 2005: The Internet of Things', (2009)

Asimakopoulou, E., Bessis, N.: Buildings and crowds: forming smart cities for more effective disaster management. In Fifth International Conference on Innovative Mobile and Internet Services in Ubiquitous Computing (IMIS), pp. 229–234 (2011)

Attwood, A., Merabti, M., Fergus, P., Abuelmaatti, O.: Sccir: Smart cities critical infrastructure response framework. In Developments in E-systems Engineering (DeSE), pp. 460–464 (2011)

Atzori, L., Iera, A., Morabito, G.: The internet of things: a survey. Comput. Netw. **54**, 2787–2805 (2010)

Berners-Lee, T., Fischetti, M.: Weaving the Web: The Original Design and Ultimate Destiny of the World Wide Web by Its Inventor. Harper, San Francisco (1999)

Bogdanov A., Knudsen L.R., Le G., Paar C., Poschmann A., Robshaw M.J.B., Seurin Y., Vikkelsoe C.: PRESENT: an ultra-lightweight block cipher, In: Proceedings of Workshop on Cryptographic Hardware and Embedded Systems (CHES). Springer (2000)

Bormann, C., Castellani, A.P., Shelby, Z.: CoAP: an application protocol for billions of tiny internet nodes. IEEE Internet Comput. **16**(2), 62–67 (2012)

Brennan, R., Wei Tai, O'Sullivan, D., Aslam, M.S., Rea, S., Pesch, D.: Open Framework Middleware for intelligent WSN topology adaption in smart buildings, Ultra Modern Telecommunications & Workshops, 2009. ICUMT'09, vol., no., pp.1,7, 12–14 Oct. 2009

Chen, H., Chou, P., Duri, S., Lei, H., Reason, J.: The design and implementation of a smart building control system, In: IEEE Int. Conf. on e-Business Engineering, pp. 255–262 (2009)

Cheshire, S., Krochmal, M.: Multicast DNS. Internet Eng. Task Force (IETF), Fremont. Request for Comments: 6762 (2013)

Cirani, S., Ferrari, G., Veltri, L.: Enforcing security mechanisms in the ip-based internet of things: an algorithmic overview. Algorithms. **6**(2), 197–226 (2013)

Compton, M., Henson, C., Lefort, L., Neuhaus, H., Sheth, A.: A survey of the semantic specification of sensors. Networks. **17**(522), (2009)

Day, J., Zimmerman, H.: The OSI reference model. Proc. IEEE. **71**(12), 1334–1340 (1983)

Deak, G., Curran, K., Condell, J.: A survey of active and passive indoor localisation systems. Comput. Commun. **35**(16), 1939–1954 (2012)

Delaney, D. T., O'Hare, G. M. P., Ruzzelli, A. G.: Evaluation of energy-efficiency in lighting systems using sensor networks. Proc. of the First ACM Workshop on Embedded Sensing Systems for Energy-Efficiency in Buildings', pp. 61–66, (2009)

D'Souza, M., Wark, T., Karunanithi, M., Ros, M.: Evaluation of realtime people tracking for indoor environments using ubiquitous motion sensors and limited wireless network infrastructure. Pervas Mob Comput. **9**(4), 498–515 (2013)

Eid, M., Liscano, R., El Saddik, A.: A universal ontology for sensor networks data. In IEEE International Conference on Computational Intelligence for Measurement Systems and Applications, 2007. CIMSA 2007, pp. 59–62 (2007)

Eisenbarth, T., Kumar, S.: A survey of lightweight-cryptography implementations. IEEE Des. Test Comp. **24**(6), 522–533 (2007)

Elkhodr, M., Shahrestani, S., Cheung, H.: A smart home application based on the Internet of Things management platform, In 2015 IEEE International Conference on Data Science and Data Intensive Systems, pp. 491–496 2015

Erickson V. L., Lin Y., Kamthe A., Brahme R., Surana A., et al.: Energy efficient building environment control strategies using real-time occupancy measurements. Proc. of the First ACM Works. on Embed. Sensing Systems for Energy-Efficiency in Buildings, pp. 19–24, (2009)

Fielding, R. T.: Architectural Styles and the Design of Network-Based Software Architectures, Ph.D. dissertation, Univ. California, Irvine, CA, USA (2000)

Filipponi, L., Vitaletti, A., Landi, G., Memeo, V., Laura, G., Pucci, P.: Smart city: an event driven architecture for monitoring public spaces with heterogeneous sensors. In Sensor Technologies and Applications (SENSORCOMM), 2010 Fourth International Conference on, pp. 281–286 (2010)

Gann, D.M., Dodgson, M., Bhardwaj, D.: Physical-digital integration in city infrastructure. IBM J. Res. Dev. **55**(1 and 2), 90–99 (2011)

Ghayvat H., Mukhopadhyay S., Gui X., Suryadevara N.: WSN- and IOT-based smart homes and their extension to smart buildings. Sensors (2015)

Hall, P., Pfeiffer, U.: Urban Future 21: A Global Agenda for Twenty-First Century Cities. Spon, London (2000)

Heath, T., Bizer, C.: Linked data: evolving the web into a global data space. Synth. Lect. Semant. Web: Theory Technol. **1**(1), 1–136 (2011)

Heer, T., Garcia-Morchon, O., Hummen, R., et al.: Security challenges in the ip-based internet of things. Wirel. Pers. Commun.: Int. J. **61**(3), 527–542 (2011)

Hernandez-Munoz, J.M., Bernat Vercher, J., Munoz, L., et al.: Smart Cities at the Forefront of the Future Internet, pp. 447–462. Springer, Heidelberg (2011)

Hui, J., Thubert, P.: 'Compression format for IPv6 datagrams over IEEE 802.15.4-based networks', IETF Request for Comments (RFC) 6282 (September) (2011)

Hunkeler, U., Truong, H.L., Stanford-Clark, A.: 'MQTT-S – a publish/subscribe protocol for wireless sensor networks', In: Proceedings of the COMSWARE, pp. 791–798 2008

Janssen, M., Charalabidis, Y., Zuiderwijk, A.: Benefits, adoption barriers and myths of open data and open government. Inf. Syst. Manag. **29**(4), 258–268 (2012)

Jara, A.J., Martinez-Julia, P., Skarmeta, A.: Light-weight multicast DNS and DNS-SD (lmDNS-SD): IPv6-based resource and service discovery for the web of things, In Proc. 6th Int. Conf. IMIS Ubiquitous Comput., pp. 731–738 (2012)

Jennings, C.: IETF Internet-Draft – Transitive Trust Enrollment for Constrained Devices (2012)

Jung, C., Shin, Y., Shin, D., Nah, Y.: Sturdy of Security Reference Model of U-City Integrated Operating Center (2009)

Kamilaris, A., Trifa, V., Pitsillides, A.: HomeWeb: an application framework for Web-based smart homes. In: 18th International Conference on. IEEE, pp. e134–e139 (2011)

Karvonen, E.: Informational Societies. Understanding the Third Industrial Revolution. Tampere University Press, Tampere (2001)

Kelly, S.D.T., Suryadevara, K., Mukhopadhyay, C.: Towards the implementation of IoT for environmental condition monitoring in homes. IEEE Sensors J. **13**(10), 3846–3853 (2013)

Klein, C., Kaefer, G.: From Smart Homes to Smart Cities: Opportunities and Challenges from an Industrial Perspective (2008).

Ko, J., et al.: Connecting low-power and lossy networks to the internet. IEEE Commun. Mag. **49**(4), 96–101 (2011)

Krochmal, M., Cheshire, S.: DNS-Based Service Discovery. Internet Eng. Task Force (IETF), Fremont. Request for Comments: 6763 (2013)

Lawton, G.: Machine-to-machine technology gears up for growth. Computer. **37**, 12–15 (2004)

Le Phuoc, D., Hauswirth, M.: In Kerry Taylor, Arun Ayyagari, David De Roure (editors) Linked open data in sensor data Mashups, Proc. of the 2nd International Workshop on Semantic Sensor Networks (SSN09), Vol-522, CEUR, 2009

Lu, J., Sookoor, T., Srinivasan, V., Gao, G., Holben, B., Stankovic, J., Field, E., Whitehouse, K.: The smart thermostat: using occupancy sensors to save energy in homes. Proc. of the 8th ACM Conference on Embd. Netw. Sensor Systems (SenSys '10), pp. 211–224 (2010)

Lüthi, Y., Meisinger, R., Wenzler, M.: Pressure sensors in the HVAC industry. In: Gassmann, O., Meixner, H., Hesse, J., Gopel, J.W., Gopel, W. (eds.) Sensor Application Vol. 2: Sensors in Intelligent Buildings, pp. 173–199. Wiley-VCH, Weinheim (2001)

Maier, M., Chowdhury, M., Rimal, B.P., Van, D.P.: The tactile internet: vision, recent progress, and open challenges. IEEE Commun. Mag. **54**(5), 138–145 (2016)

McDaniel, P., McLaughlin, S.: Security and privacy challenges in the smart grid. Secur. Priv, IEEE. **7**(3), 75–77 (2009)

Metje, N., Chapman, D.N., Cheneler, D., Ward, M., Thomas, A.M.: Smart pipes-instrumented water pipes, can this be made a reality? Sensors. **11**(8), 7455–7475 (2011)

Milton, N., P. S. e. a.: Combining cloud and sensors in a smart city environment. EURASIP J Wirel Commun Netw, p. 247 (2012)

Montenegro, G., Kushalnagar, N., Hui, J., Culler, D.: 'Transmission of IPv6 packets over IEEE 802.15.4 networks', IETF Request for Comments (RFC) 4944 (September) (2007)

Montenegro, G., Kushalnagar, N., Hui, J., Culler, D.: 'Transmission of IPv6 Packets over IEEE 802.15.4 Networks,' RFC 4944, Internet Engineering Task Force, 2007 (2010)

MQ Telemetry Transport.: http://mqtt.org. Accessed 10 Mar 2015 (n.d.)

Mulligan, G.: 'The 6LoWPAN architecture', EmNets'07: Proceedings of the 4th workshop on Embedded networked sensors, ACM (2007)

Myers, C.: Intelligent Buildings: A Guide for Facility Managers. UpWord Publishing, New York (1996)

Nasir, A., Hussain, S., Soong, B., Qaraqe, K.: Energy efficient cooperation in underlay RFID cognitive networks for a water smart home. Sensors. 2015

Nikander, P., Gurtov, A., Henderson, T.R.: Host identity protocol (HIP): connectivity, mobility, multi-homing, security, and privacy over IPv4 and IPv6 networks. IEEE Commun Surv Tut. **12**(2), 186–204 (2010)

Nye, D.: American Technological Sublime. MIT Press, Cambridge, MA (1996)

Ratti, C., Townsend, A.: Harnessing Residents' Electronic Devices Will Yield Truly Smart Cities (2011)

Shadbolt, N., O'Hara, K., Berners-Lee, T., Ibbins, N., Glaser, H., Hall, W., Schraefel, M.C.: Linked open goverment data: lessons from Data.gov.uk. IEEE Intell. Syst. **27**(3, Spring Issue), 16–24 (2012). doi:10.1109/MIS.2012.23

Shelby, Z., Hartke, K., Bormann, C., Frank, B.: 'Constrained Application Protocol (CoAP)', draft-ietf-core-coap-11 (2012)

Shi, W., Dustdar, S.: The promise of edge computing. Computer. **49**(5), 78–81 (2016)

Simsek, M., Aijaz, A., Dohler, M., Sachs, J., Fettweis, G.: 5G-enabled tactile internet. IEEE J Sel Areas Commun. **34**(3), 460–473 (2016)

Singhvi, V., Krause, A., Guestrin, C., Garrett, J.H. Jr., Matthews, H.S.: Intelligent light control using sensor networks, In: Proc. of the 3rd International Conference on Embedded Networked Sensor Systems, SenSys '05, ACM, NY, USA, pp. 218–229 (2005)

Skouby, K.E., Kivimäki, A., Haukiputo, L., Lynggaard, P., Windekilde, I.: Smart Cities and the Ageing Population; Proceedings of the 32nd Meeting of WWRF, 2014

So, A.T.P., Tse, B.W.L.: Intelligent air-conditioning control. In: Gassmann, O., Meixner, H., Hesse, J., Gardner, J.W., Gopel, W. (eds.) Sensor Application Vol. 2: Sensors in Intelligent Buildings, pp. 29–61. Wiley-VCH, Weinheim (2001)

Standaert, F.-X., Piret, G., Gershenfeld, N., Quisquater, J.-J.: SEA: a scalable encryption algorithm for small embedded applications, In: Proceedings of 7th IFIP WG 8.8/11.2 CARDIS, pp. 222–236. Springer, 2006

Suzuki, L C S R: Data as Infrastructure for Smart Cities. PhD Thesis. University College London, UK (2015a)

Suzuki, L. C.D.S.: The internet of things, the hype and the Road map, IET Eng Technol Ref., pp. 1–12 (2015b)

Terzis, S., Nixon, P., Narasimhan, N., Walsh, T.: Middleware for pervasive and ad hoc computing. Pers. Ubiquit. Comput. **10**(1), 4–6 (2005)

Tiberi, U.; Fischione, C.; Johansson, K.H.; Di Benedetto, M.D.: 'Adaptive self-triggered control over IEEE 802.15.4 networks,' Decision and Control (CDC), 2010 49th IEEE Conference on, vol., no., pp.2099, 2104, 15–17 Dec. 2010

Trankler, H.R., Kanoun, O.: Sensor systems in intelligent buildings. In: Gassmann, O., Meixner, H., Hesse, J., Gopel, J.W., Gopel, W. (eds.) Sensor Application Vol. 2: Sensors in Intelligent Buildings, pp. 485–510. Wiley-VCH, Weinheim (2001)

van den Besselaar, P., Beckers, D.: The Life and Death of the Great Amsterdam Digital City Digital Cities III, pp. 1–16. Springer, Berlin (2005)

Wang, S.W., Xie, J.L.: Integrating building management system and facility management on internet. Autom. Constr. **11**(6), 707–715 (2002)

Wang, W., Dong, L., Wang, C.: A Linked-Data Model for Semantic Sensor Streams, Green Computing and Communications, 2013 IEEE and Internet of Things, pp. 468–475 (2013)

Wen, Y., Yang, X., Xu, Y.: Cloud-Computing-Based Framework for Multicamera Topology Inference in Smart City Sensing System. (2010)

Shelby Zach, Bormann Carsten: '6LoWPAN: The Wireless Embedded Internet', Wiley Publishing (2009)

Zhang, Q., Cheng, L., Boutaba, R.: Cloud computing: state-of-the-art and research challenges. J. Internet Serv. Appl. **1**, 7–18 (2010)

Zhang, S., McCullagh, P., Nugent, C., Zheng, H., Black, N.: An ontological framework for activity monitoring and reminder reasoning in an assisted environment. J. Ambient. Intell. Humaniz. Comput. **4**(2), 157–168 (2013)

Cities as Visuospatial Networks

A. Natapov and D. Fisher-Gewirtzman

Abstract Current methods used in the study of urban systems are based mostly on economic and transportation demands and ignore human spatial cognition processes, like visual perception, while cognition is an active player in the evolution and dynamics of urban space.

This chapter presents a collection of interdisciplinary studies that link visuospatial cognition to urban dynamics. These are located at the intersection of three rapidly developing scientific, technological, and practical fields: spatial cognition (acquisition and utilization of spatial knowledge), complexity science (graph and network theories), and smart cities (urban planning and design-enhancing digital technologies). We use two complementary methods: (1) Spatial graph-based analysis – Urban environment is represented as a chain of navigational decisions in a form of mathematical graphs (or networks). Then several centrality measures from the graph theory are applied to the constructed graphs to evaluate structural position of urban locations. (2) Computational simulation of the visual search: Pedestrian visual search for urban locations is conceptualized as a stochastic process and modeled by random walk simulation. Results of the simulation are used to quantify visual accessibility of diverse urban settings.

Taken together, suggested methods construct a novel cognitive paradigm in the study of urban systems and urban modeling. The results of the studies present effective tools for exploring various scenarios of urban sustainable design, reshaping infrastructure, mobility, and architecture of our cities.

A. Natapov (✉)
Centre for Advanced Spatial Analysis, University College London, 90 Tottenham Court Road, W1T 4TJ, London, UK
e-mail: a.natapov@ucl.ac.uk

D. Fisher-Gewirtzman
Faculty of Architecture & Town Planning, Technion-IIT, Israel, Sego building, Technion city, Haifa 32000, Israel

S.T. Rassia, P.M. Pardalos (eds.), *Smart City Networks*, Springer Optimization and Its Applications 125, DOI 10.1007/978-3-319-61313-0_11

1 Linking Urban Space, Activities, and Walking

Given the unprecedented urbanization, urban planners and designers need models to support sustainable planning and design, especially a design for walking, the most environmentally, socially, and economically sustainable mode of transport. Walking routines depend on distribution of various activities in space, while the spatial arrangement of socioeconomic activities is shaped by the interaction of many entities in the context of a particular urban structure. With economic, social, and transportation factors, cognitive properties of city dwellers, particularly their visual ability, are an active player in the evolution of urban activities.

Commonly, movement-seeking urban uses and activities, such as shops, cafés, restaurants, and galleries, are clustered in certain areas of the city, which is a spontaneous, bottom-up urban phenomenon. It is explained by the underlying financial logic of the clustering that increases patronage, magnifies opportunities, and economizes travel costs (Huang and Levinson 2011). However, these explanations are based on economic principles alone and are not concerned with the cognitive characteristics of city dwellers, pedestrians.

In urban studies, two approaches attempt to explain interaction between pedestrian movement and urban activity distribution. The first, the attraction approach, focuses on the spatial distribution of activities (Timmermans et al. 2002; Wilson 2000), while the second, configurational approach, attributes the choice of routes relative to the morphology of the street network, a continuous network of passable open spaces (Hillier 1996).

The attraction approach focuses on the spatial distribution of urban activities within the buildings as pedestrian trip generators and as the locations to which pedestrians are being attracted (e.g., Pushkarev and Zupan 1975). The fundamental characteristic of such models is the route choice as a result of minimization of travel time and distance. The outcome is a single travel path from among all possible paths the urban morphology provides. The main concern with many of these models has been their unrealistic behavioral assumptions. Foremost among these has been the assumption of utility maximization, which has allowed the development of models in an optimization framework.

Configurational approach emerged in the field of architectural research. It attributes urban movement and functional development to the configuration of passable open spaces and, in particular, their visibility caused by the urban layout (Hillier et al. 1993; Hillier and Hanson 1984). The term, configuration, refers to the way every space in the environment relates to every other. According to this approach, the layout of the environment affects the choices individuals make.

The configurational approach considers wayfinding behavior, which is not rigidly constrained and can be more adventuresome and exploratory (Montello 2001; Montello and Pick 1993; Weisman 1981; O'Neill 1991). It is cognitively meaningful, since it uses the human sense of sight as a generative factor to construct the spatial representation. Unlike the simple notation of traveled metric distance used in the attraction theory, such cognitive approach reflects more complex behavioral factors involved in navigating to a place.

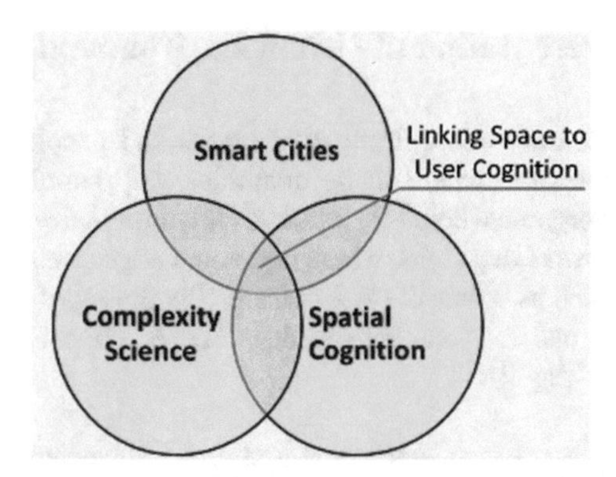

Fig. 1 Conceptual framework of the intersectional aspects involved in the research

Both attraction and configurational approaches are far from providing a complete account of urban dynamics and do not portray the entire picture of urban domain. The configurational approach considers only street configuration and totally ignores urban uses and activities, while the attraction approach deals only with attractors, disregarding the impact produced by urban morphology.

1.1 Visuospatial Cognition and Built Environment

In this chapter, we link built environment and the way people use it by discovering how urban visibility contributes to the generation of activity patterns and how these are related to pedestrian movement. We address three fundamental questions: (1) How could visual structure and visual connectivity of socioeconomic activities in the city be quantified? (2) What are the relations of visual proximity of street network and urban activity locations? and (3) How could pedestrian visual search for urban locations be simulated and predicted?

This research presents an interdisciplinary study at the intersection of three rapidly developing scientific, technological, and practical fields: spatial cognition (acquisition and utilization of spatial knowledge), complexity science (statistical physics, graph, and network theories), and smart cities (a new and promising direction in urban planning and design-enhancing digital technologies) (Fig. 1).

Taken together, these research steps construct a novel urban-behavioral paradigm. This innovative paradigm will stimulate a collection of methods to answer questions as to how relationships between cognitive and physical objects comprise our cities.

2 Graph Representation of Urban Environment

Urban areas are not only concentrations of places and people but also "systems of organized complexity" where many quantities vary simultaneously and "are interrelated into an organic whole" (Jacobs 1961). This organized complexity can be represented by networks or graphs, where nodes and edges are embedded in space.

A graph is denoted as a pair G (V, E) where V is the set of vertices (or nodes), $V = \{v_1, v_2, \ldots v_n\}$, and E is the set of edges (1). A graph is represented by an adjacency matrix A (Fig. 2):

$$A_{\mathrm{ij}} = \begin{cases} 1, \text{the vertices } v_i \text{ and } v_j \text{ are connected,} \\ 0, \text{otherwise.} \end{cases} \quad (1)$$

Graph methods allow combining behavioral and spatial features in one quantitative framework. Recently, there has been a growing overlap between cognitive science and spatial modeling, using vision and visibility as generative factors in understanding how spatial properties of the built environment affect human mobility (Natapov and Fisher-Gewirtzman 2016; Hölscher et al. 2006; Wiener and Franz 2005). For instance, the axial graph method developed in Space Syntax, a set of configurational tools for urban design, produces a graph of straight street segments traced over each of the longest sightlines (Hillier and Hanson 1984). The segments are linked into a network via their intersections and analyzed as a network of movement options.

Another vision-based graph approach is a visibility graph applied to the urban environment by imposing a regular or an irregular grid on the top of urban space. Such a graph is created in terms of how each point of the grid is visible to others. Visibility graph applications differ, regarding implementation; some analyze the open space between the built forms, while others (Krüger 1979; De Floriani

Node	1	2	3	4
1	0	1	1	1
2	1	0	1	0
3	1	1	0	0
4	1	0	0	0

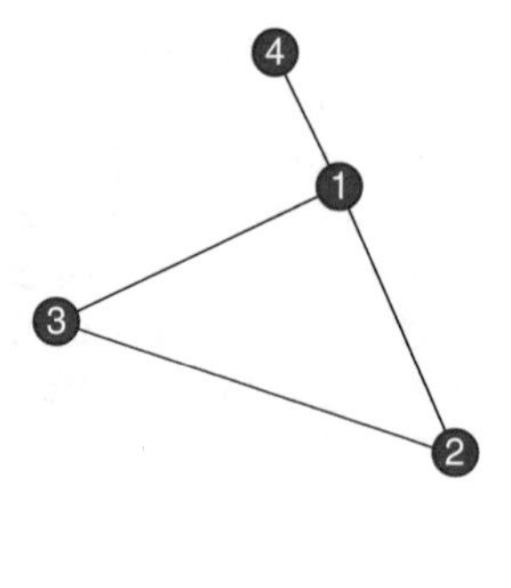

Fig. 2 An exemplary graph and its adjacency matrix

et al. 1994; Fisher-Gewirtzman and Natapov 2014) map the intervisibility of built components. Turner et al. (2001) and Turner and Penn (1999) applied a visibility graph to the analysis of architectural space, taking a grid of all existing points across the space, rather than selecting a few key locations.

Jiang and Claramunt) 2002) adopted the visibility graph, relying on a set of characteristic points in urban layout. The characteristic points are defined as the nodes of the urban structure and visibility connections between them as edges schematized as a graph. This method has several advantages over other graph representations: It is computable and cognitively meaningful. It is easier to understand, and it provides a much richer interpretation of the space syntax (Batty 2004).

2.1 An Alternative Representation: Integrative Visibility Graph (IVG)

Existing graph methods used in the analysis of urban environments consider only street network and ignore the functional aspect of the built form, whereas functions or urban activities are extremely important features of urban landscape.

Based on Jiang and Claramunt (2002), we introduce an Integrative Visibility Graph (IVG), a novel method to encode urban space, urban functions, and their mutual visibility properties in one graph representation (Natapov et al. 2013). Nodes in the graph stand for two types of decision points along the traveler's path: first, the intersections and turning points within the street network, meaning those places where a traveler must make navigational decisions; and second, points of interest, namely, the locations of urban activities. The nodes are connected if they are mutually visible (Fig. 3).

A path in the graph is a hypothetical visual route taken by a traveler searching for a specific place, function, or location in the city.

The constructed graph accounts for the exact location of the particular building or building entrance throughout the visibility network. It also enables separating the navigational decision points within the street network (intersections) from the functional targets (urban activities), neither of which is addressed in most urban graph representations.

This method links the city's functional structure to spatial cognition by focusing on one basic cognitive property involved in urban wayfinding, human vision. The main innovation of the proposed representation method is an ability to integrate three different, but interrelated, factors of urban landscape – structure, functioning, and user behavior. In contrast to other graph representations, our graph is deterministic, computable, and cognitively meaningful.

The proposed Integrative Visibility Graph enables combining graph-based measures with more conventional functionality of GIS analysis. Since this structural analysis is based on individual point values, additional external socioeconomic data, which are mostly point based, can be used for further analysis. The developed graph method also allows incorporating metric distance in one framework with topological properties.

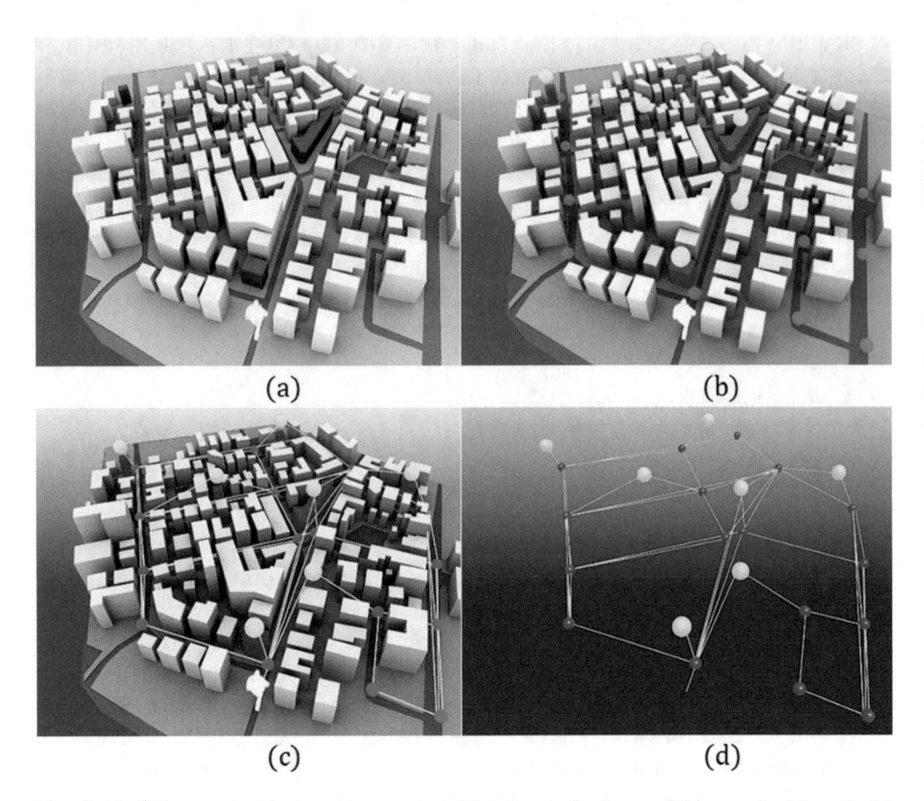

Fig. 3 Visibility graph: (**a**) open city space and locations of urban activities marked in *red*; (**b**) decision points – intersection in *blue* dots and activity locations in *yellow* dots; (**c**) *lines* of sight between decision points; (**d**) resulting visibility graph of structural and functional locations

2.2 Accessibility of the Network

We study one predefined type of urban functions or activities, food and drink public facilities. These facilities, such as cafés, coffee shops, restaurants, and others, are known as "third place" in urban sociology and play an important role in establishing a sense of place. Ray Oldenburg, an urban sociologist, author of the books *Celebrating the Third Place* and *The Great Good Place* (1989), has developed a concept of third place related to informal public-gathering places. The term refers to social surroundings separated from the two typical social environments: home and workplace. These informal meeting places are a new trend of modern cities, and a community is intentionally seeking them as vital to current societal needs.

The proposed graph representation allows us to compute measures of visual access to particular points of the urban layout by implementation of the graph and network theoretical methods. In Natapov et al. (2013), we carry out network structural analysis, based on the centrality model. This model was first developed and used in social network studies (Freeman 1979). We examine three centrality measures – degree centrality, closeness centrality, and betweenness centrality applied to the IVG.

The degree centrality measures the number of nodes that interconnect a given node. The degree measure captures how many destinations can be seen from each node within given geometrical conditions. The degree $C_D(v_i)$ of the node v_i is defined by:

$$C_D(v_i) = \#\{j \in V : \text{vertices } v_i \text{ and } v_j \text{ are connected}\} \tag{2}$$

The closeness centrality measures how many steps are required to access every other node from a given node. The normalized closeness centrality is defined by the inverse of the average length of the shortest paths to all the other nodes in the graph:

$$C_C(v_i) = \frac{n-1}{\sum_{k=1}^{n} d(v_i, v_k)} \tag{3}$$

where $d(v_i, v_k)$ is the topological distance between v_i and v_k. This measure determines the relative importance of a node within the graph. The more integrated nodes with short distances to others have bigger closeness centrality. It is important to note that degree centrality shows how much one can directly see from one point location, while closeness centrality shows how much one can see, both directly and indirectly, from one point location. For example, one location may not be directly visible from another, but it may be visible via a third location; then we say they are indirectly visible through an intermediate location.

The betweenness centrality is another centrality measure of a node within a graph. It captures how often, on average, a location may be used in journeys from all places to all others. Locations that occur on many of the shortest paths between others have higher betweenness than those that do not. Betweenness centrality is estimated as follows:

$$C_B(v_k) = \sum_i \sum_j \frac{P_{ikj}}{P_{ij}} \tag{4}$$

where P_{ij} = # of shortest paths from v_i to v_j, P_{ikj} = # of shortest paths from v_i to v_j through v_k.

Betweenness centrality measures the influence a node has over the flow through the network. A node with a high degree of centrality does not guarantee it is well connected to all other nodes. Sometimes, a node with a few direct connections is more important, since it can act as a bridge, meaning that, without it, a network may be broken into two or more subgraphs, and this property is controlled by betweenness centrality (Fig. 4).

We calculate centrality measures for the IVG and SNVG and use them to analyze the relations between urban activity locations and street configuration. The found correlation assures the use of visual accessibility as a linkage between street

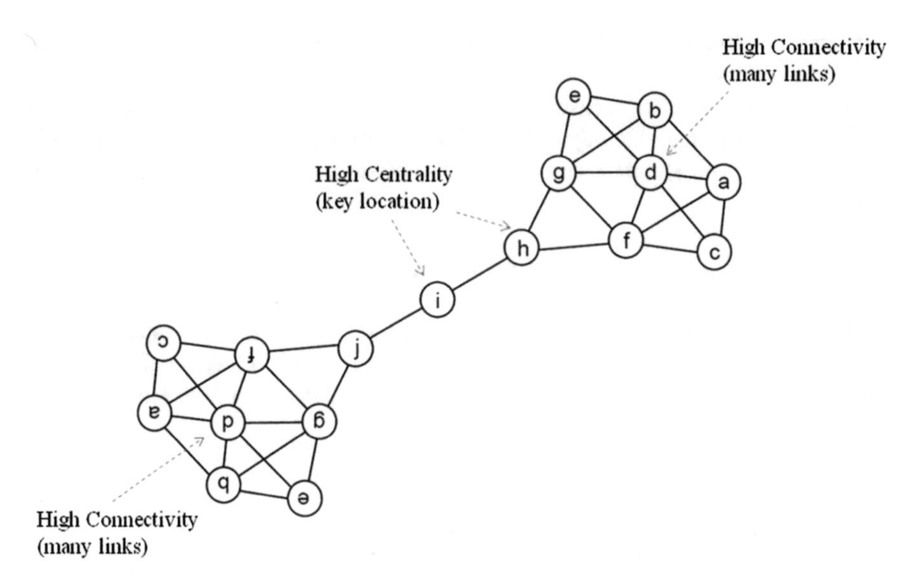

Fig. 4 Different types of centrality. Node i has high betweenness value and low degree value, while node d and p are the opposite

configuration and activity location and determines whether visibility of urban layout is among predictors of successful urban retail location.

2.3 *Efficiency of the Centrality Measures*

It has been found among Space Syntax researchers that pedestrian movement rates are significantly correlated with the local centrality values, and highly integrated streets attract more people than segregated ones (Hillier et al. 1993). Each of the three graph-theoretic measures we examined has a different meaning and assigns different interpretation to urban interplay. The measures of degree and closeness centralities reveal the set of most visually connected streets and third places, while betweenness centrality shows the most visually attracted locations among the whole set of existing visual opportunities of the city. Figure 5 shows most visually connected streets in the case study, the historical district of Tel Aviv-Yafo. These are Allenby and Jaffa Streets and Rothschild Boulevard.

Our findings reveal that cafés, coffee shops, restaurants, and other food and drink facilities are positively related to all three centrality measures. The correlation with the closeness centrality, indicating the visual accessibility of the node in the system, is very strong. Urban activities located in the least visually integrated venues are more likely to be less popular and vice versa. We conclude that, among the centrality measures, the closeness centrality is the best predictor of "third place" distribution.

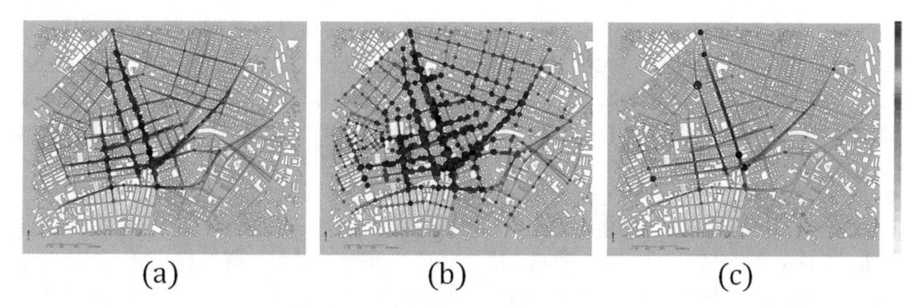

Fig. 5 Distribution of (**a**) visibility degree, (**b**) closeness centrality, (**c**) betweenness centrality with IVG. *Red* dots are third place locations, while *blue* ones are junctions and turning points. Each dot size represents the magnitude of the measures

A traditional view in city planning theory is that services and shops of different levels should be consistently located in places of a corresponding level of importance (Ben-Joseph 2005; Osborn et al. 1963). According to this concept, major urban streets should be attractors for global urban functions, while local services, such as cafe shops and restaurants, could be accommodated along secondary, quiet streets. Our results have demonstrated the opposite – majority of third place activities are located on the major urban roads not within the secondary areas.

3 Visual Search for Urban Locations

Urban spatial structure results from multitudes of decisions by heterogeneous urban agents, such as consumers of housing and urban amenities, land developers, commuters, and pedestrians (Batty 2013; Zachariadis 2005). Computer modeling, namely, agent-based modeling (ABM), can imitate this bottom-up phenomena and serves for testing of different hypotheses and theories of urban change (Rose and Ligtenberg 2014; Benenson et al. 2009; Benenson 1998). In Natapov et al. (2016), we simulate and analyze a search within an urban environment as a stochastic (i.e., random) process with the use of ABM.

Random walks, sometimes called drunkards' walks, first introduced by Karl Pearson in 1905 (Pearson 1905), are a mathematical formalization of a path comprising a succession of random steps. Random walks represent a natural stochastic process of diffusion, in our case, a diffusion of visual scanning in the urban environment. Instead of describing a process that can only evolve in one way, the stochastic process contains indeterminacy, meaning there are infinite directions in which the process may evolve.

Examples of random walk implementations are found in urban studies and studies of visual cognition, both fields involved in our work. Volchenkov and Blanchard explore street networks of several cities by random walks (Blanchard and Volchenkov 2008, 2009; Volchenkov and Blanchard 2007). Batty (2003)

investigates the influence of randomness and geometry on movement and locational patterns in agent-based pedestrian models. Jiang and Jia use random agents to simulate human movement in large street networks. They found that, given a street network, movement patterns generated by purposive walkers and by random walkers are the same (Jiang and Jia 2009). In cognitive science, several random walk methods were developed for modeling visual search and eye movement. Brockmann and Geisel (1999) introduced a phenomenological model for generating human visual scan paths and found that such optimization implies scan-paths are geometrically similar to random walks. Boccignone and Ferraro (2004) modeled gaze shifts – eye movements significant in the visual information selection process by means of random walk simulation.

3.1 Random Walks on the Visibility Network

Although a scene or a walk in the city is not randomly viewed, but influenced by environmental and behavioral factors, the aggregated visual search can be represented stochastically. Here, visual search is modeled as a memoryless Markov chain, where the next step of an agent does not depend on its previous step. At each step, the agent chooses the next node in the graph from the neighbors of the current node with equal probability. The transition probability, the probability of walking from node v_i to node v_j is therefore:

$$p_{ij} = \begin{cases} \dfrac{1}{\text{degree } of\ the\ node\ v_i} & if\ the\ nodes\ v_i\ and\ v_j\ are\ connected \\ 0, otherwise \end{cases} \tag{5}$$

Random agents are used to approximate the first passage times (FPT) or the first hitting times to nodes in the graph. For any node v_i, the FPT n_i is the expected number of steps required for the random walker to reach the node for the first time, starting from a node randomly chosen among all nodes of the graph G (Blanchard and Volchenkov 2009). This characteristic time is calculated as an average over the lengths of all the paths toward the node taken in accordance with their respective probabilities.

Recurrence time (RT) indicates how long a random walker is expected to wander before revisiting the node. In the case of random walks, it can be calculated from the stationary probability (see e.g., Blanchard and Volchenkov 2008). Random walks determine a unique stationary probability on the graph – the probability of finding a random walker at a certain node after a long enough run. For an undirected graph, the stationary probability p_i of the node v_i is defined as

$$p_i = \frac{k_i}{\sum_{v_i \in V} k_i}, \text{where } k_i = C_D(v_i) \tag{6}$$

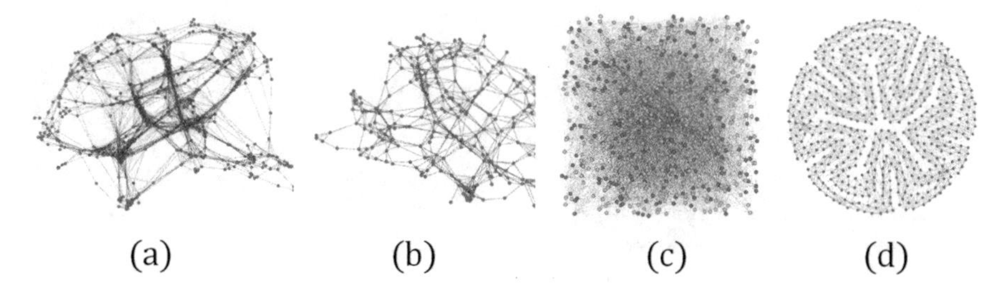

Fig. 6 Non-geo-referenced visualizations of the visibility graphs: (**a**) actual functional VG, (**b**) actual street VG, (**c**) simulated random VG, (**d**) simulated regular VG

The expected recurrence time, m_i to $v_i \in V$, is the inverse of stationary probability distribution of random walks as expressed by

$$m_i = p_i^{-1} \tag{7}$$

3.2 Simulation on the Visibility Graphs

Random agents wandering on the graph imitate aggregated visual search for urban activities in the case study, the historical district of Tel Aviv-Yafo (Natapov et al. 2016). Then the case study is compared to a set of systematically generated computational patterns represented by various visibility graph models (Fig. 6).

A comparison between local visibility properties, estimated by RT, and global visibility properties, estimated by FPT, is carried out. A positive global-local relationship of spatial networks is termed intelligibility in urban studies and is deemed a key determinant in human navigation and wayfinding (Hillier 1996). Additionally, we compare the first passage time, an outcome of stochastic simulation, with deterministic graph theoretic measures of centrality, namely, closeness and betweenness centralities.

Figure 7 shows fragments of different urban tissues from cities across the globe. These are possible examples of regular and random environments. For instance, Barcelona and San Francisco (Fig. 7a, b) are based on the orthogonal regular street pattern and could be referred as the regular graph, while Ahmedabad and Venice (Fig. 7c, d) have random, irregular patterns. These cities have different morphology and different historical roots; Barcelona and San Francisco are relatively modern, planned cities, while Ahmedabad and Venice are older and evolved organically.

Comparison of distinct visibility types aims to identify a navigational signature of the case study and allows measurement of visual distinction between different locations in the city. Our main results disclose a strong positive relation between local and global visibility properties.

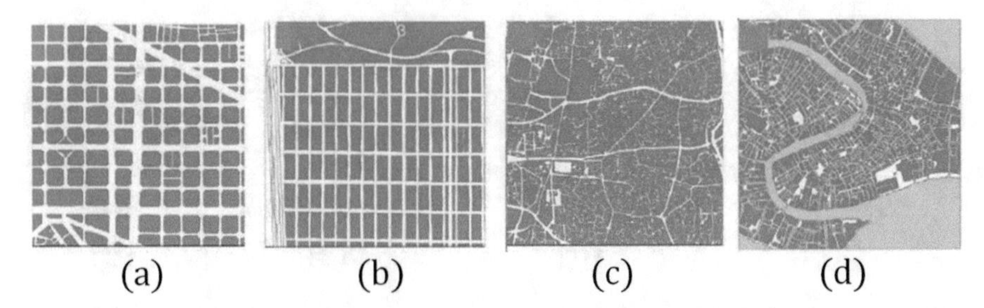

Fig. 7 Urban tissues (**a**, **b**) illustrate regular layouts, (**c**, **d**) random layouts: (**a**) Barcelona, (**b**) San Francisco, (**c**) Ahmedabad, (**d**) Venice (Drawings from Jacobs 1993)

Though the movement of random walkers does not describe the actual visual process and actual pedestrian movement in the strict sense, it concerns topological and geometric properties of the environment and gives a sense of the notion of visuospatial search. Random agents reveal the overall connectivity of the environment and assist in estimating the probable access times to the "ambiguous" locations of the urban environment.

4 Further Directions for Cognitive Modeling

Presented studies reveal several directions for future methodological improvements and potential contribution to the urban design field. These are summarized as follows:

A computational model that outlines growth and evolution of urban activities could be designed in future research. Encoding cognitive properties of urban setting as a graph reflects a fundamental network structure of urban environment and allows it to be viewed as a growing visibility network of interacting agents and activities. The model is intended to simulate interactions between cities' spatial and cognitive features, assessing their effects on the system as a whole; that is, how individuals interact with urban environment and how the environment constantly evolves and transforms, bottom-up, to generate new urban activity locations.

The model might be constructed, based on the dynamic network approach (DNA), an emergent scientific field in complexity science that brings together traditional network analysis, simulation, and multi-agent systems within network theory (Newmaern 2003; Barabási and Albert 1999). The modeling could reflect overall network organization, namely, distribution of visibility lines. Visibility distribution is an outcome of growth dynamics, within a network space, and uncovers the rules behind efficiency and robustness of the successful urban environment. It could be described as a "preferential attachment" scenario (Barabási and Albert 1999). Here, visibility lines are distributed among several navigational locations, according to how many they already have, so those already "rich" receive more than those who are not. Finally, the model could be statistically evaluated against the case study dataset of activity locations. The simulation will contribute to development of the design tool to be used along planning and design processes in new, rapidly urbanizing contexts.

Models that connect spatial behavior to urban development have become more dominant over the past 30 years (McFadden 2000). However, ways to incorporate spatial cognition in such models have not yet been widely adopted by design professionals, presumably because of difficulties in operationalizing and measuring constructs, such as visibility, spatial memory, and wayfinding. The research described in this chapter set out to bridge the gap, linking visual perception and movement in urban context with graph measures and physics-based computational simulations. It connects concepts, like intelligibility and visibility, originating in architectural studies, and the random walks concept used in the graph-theoretical research. Analyzing both behavioral and morphological properties of the city makes it possible to predict urban activity traits.

The studies presented in this chapter are drawn on complexity and network theories. They show that urban dynamics are related to complex network theory and test a new probable link between urban-social processes and physical theories. Most complex science applications study how individual parts interact to produce global behavior in telecommunication, biology, transport, or social systems. Here, we introduce a new type of network, urban visibility network, which could open a new direction in complexity applications.

To conclude, we focus on the behavioral properties of walking, the most environmentally, socially, and economically sustainable of all urban transport modes. How friendly our cities are to walking affects many important social indicators, such as household travel behavior, energy expenditure, and public health. These outcomes are influenced by urban form and land use. A higher concentration of nonwork land uses appears to reduce vehicle miles traveled (Krizek 2003), to decrease urban energy consumption (Newman and Kenworthy 1999), to produce better health indicators, and to foster social cohesion (Jacobs 1961; Pendola and Gen 2008).

Initial step for planners dealing with urban use development should understand the importance of the human factor in mechanisms of the urban realm. More attention should be paid, not to creating new establishments but to understanding conditions that allow these functions to emerge and to thrive on their own. Particularly, street-fronting commercial uses should be allocated along major streets. We see this in the historic districts: high streets, souks, boulevards – all are packed with street-fronting uses. Turning the city's main streets over to the car and withdrawing commercial uses to the secondary streets – as has happened as planned, modern cities worldwide – goes against the grain of urban evolution.

Studies presented in this chapter offer methodology for understanding the relationships between the built environment and our cognitive functions while showing how professionals can use research in spatial cognition to design better spaces and cities. The studies' outcomes could be integrated into practical tools for urban planning and design. Such tools will lead to socioeconomic impact on built environment, supporting our society in preventing car-dependent sprawls and improving public health and mobility.

Acknowledgements We thank Prof. Daniel Czamanski for his constructive comments and Dr. Michael Natapov for his valuable contribution in development and implementation of the visibility algorithm and the computational simulation.

References

Barabási, A.L., Albert, R.: Emergence of scaling in random networks. Science. **286**(5439), 509–512 (1999)
Batty, M.: Agent-based pedestrian modelling. In: Longley, P.A., Batty, M. (eds.) Advanced Spatial Analysis: the CASA Book of GIS, pp. 81–108. ESRI Press, Redlands (2003)
Batty, M.: A new theory of space syntax. Syntax. **44**(0), 36 (2004)
Batty, M.: The new Science of Cities. The MIT Press, Cambridge, MA (2013)
Benenson, I.: Multi-agent simulations of residential dynamics in the city. Comput. Environ. Urban. Syst. **22**(1), 25–42 (1998)
Benenson, I., Hatna, E., Or, E.: From schelling to spatially explicit modeling of urban ethnic and economic residential dynamics. Sociol. Methods Res. **37**, 463–497 (2009)
Ben-Joseph, E.: The Code of the City: Standards and the Hidden Language of Place Making. MIT Press, Boston (2005)
Blanchard, P., Volchenkov, D.: Intelligibility and first passage times in complex urban networks. Proc. R Soc A Math. Phys. Eng. Sci. **464**, 2153–2167 (2008)
Blanchard, P., Volchenkov, D.: Mathematical Analysis of Urban Spatial Networks. Springer, Berlin (2009)
Boccignone, G., Ferraro, M.: Modelling gaze shift as a constrained random walk. Physica A. **331**, 207–218 (2004)
Brockmann, D., Geisel, T.: Are Human Scanpaths Levy Flights? Artificial Networks, Conference Publication No. 470, IEE, pp. 263–268. (1999)
De Floriani, L., Marzano, P., Puppo, E.: Line-of-sight communication on terrain models. Int. J. Geogr. Inf. Syst. **8**, 329–342 (1994)
Fisher-Gewirtzman, D., Natapov, A.: Different approaches of visibility analyses; applied on a hilly urban environment. Surv. Rev. **46**(338), 366–382 (2014)
Freeman, L.C.: Centrality in social networks: conceptual clarification. Soc. Networks. **1**, 215–239 (1979)
Hillier, B.: Space is the Machine: A Configurational Theory of Architecture. Cambridge University Press, Cambridge (1996)
Hillier, B., Hanson, J.: The Social Logic of Space. Cambridge University Press, Cambridge (1984)
Hillier, B., Penn, A., Hanson, J., Grajewski, T., Xu, J.: Natural movement: or, configuration and attraction in urban pedestrian movement. Environ. Plann. B Plann. Des. **20**, 29–66 (1993)
Hölscher, C., Meilinger, T., Vrachliotis, G., Brösamle, M., Knauff, M.: Up the down staircase: wayfinding strategies and multi-level buildings. J. Environ. Psychol. **26**(4), 284–299 (2006)
Huang, A., Levinson, D.: Why retails cluster: an agent model of location choice on supply chains. Environ. Plan. B: Plan. Des. **38**(1), 82–94 (2011)
Jacobs, J.: The Death and Life of Great American Cities, Vintage Books edn. Vintage Books, New York (1961)
Jacobs, A.: Great Streets. MIT Press, Boston (1993)
Jiang, B., Claramunt, C.: Integration of space syntax into GIS: new perspectives for urban morphology. Trans. GIS. **6**(3), 295–309 (2002)

Jiang, B., Jia, T.: Agent-based simulation of human movement shaped by the underlying street structure. Technology. **1**, 1–2 (2009)
Krizek, K.J.: Residential relocation and change in urban travel: does neighborhood-scale urban form matter? J. Am. Plan. Assoc. **69**(3), 265–279 (2003)
Krüger, M.T.J.: An approach to built-form connectivity at an urban scale: system description and its representation. Environ. Plan B: Plan. Des. **6**, 67–88 (1979)
McFadden, D.L.: Disaggregate Travel Demand's RUM Side: a 30-Year Retrospective (Manuscript). Department of Economics, University of California, Berkeley (2000)
Montello, D.: Spatial cognition. In: Smelser, N.J., Baltes, P.B. (eds.) International Encyclopedia of the Social & Behavioral Sciences. Elsevier, Amsterdam (2001)
Montello, D., Pick, H.: Integrating knowledge of vertically aligned large-scale spaces. Environ. Behav. **25**(4), 457–484 (1993)
Natapov, A., Czamanski, D., Fisher-Gewirtzman, D.: Can visibility predict location? Visibility graph of food and drink facilities in the city. Surv. Rev. **45**(333), 462–471 (2013)
Natapov, A., Czamanski, D., Fisher-Gewirtzman, D.: Visuospatial search in urban environment simulated by random walks. Int. J. Des. Creat. Innov. **4**(2), 85–104 (2016)
Natapov, A., Fisher-Gewirtzman, D.: Visibility of urban activities and pedestrian routes: an experiment in a virtual environment. Comput. Environ. Urban. Syst. **58**, 60–70 (2016)
Newmaern, M.E.J.: The structure and function of complex networks. SIAM Rev. **45**, 167–256 (2003)
Newman, P., Kenworthy, J.: Sustainability and Cities: Overcoming Automobile Dependence. Island Press, Washing, DC (1999)
Oldenburg R: Great Good Place; Cafés, Coffe Shops, Bookstores, Bars, Hair Salons, and Other Hangouts at the Heart of a Community. Marlow & Company, New York (1989)
O'Neill, M.: Evaluation of a conceptual model of architectural legibility. Environ. Behav. **23**(3), 553–574 (1991)
Osborn, F., Mumford, L., Whittick, A.: The new Towns: the Answer to Megalopolis. McGraw-Hill, New York (1963)
Pearson, K.: The problem of the random walk. Nature. **72**, 294 (1905)
Pendola, R., Gen, S.: Does "main street" promote sense of community? A comparison of San Francisco neighborhoods. Environ. Behav. **40**(4), 545–574 (2008)
Pushkarev, B., Zupan, J.M.: Urban Space for Pedestrians. MIT, Cambridge (1975)
Rose, J., Ligtenberg, A. L.: Simulating Pedestrians Through the Inner-City: An Agent-Based Approach. Proceedings of the Social Simulation Conference (SSC'14). Barcelona. (2014)
Timmermans, H.J.P., Arenze, T., Joh, C.-H.: Analyzing space-time behavior: new approaches to old problems. Prog. Hum. Geogr. **26**(2), 175–190 (2002)
Turner, A., Doxa, M., O'Sullivan, D., Pen, A.: From isovists to visibility graphs: a methodology for the analysis of architectural space. Environ. Plan. B: Plan. Des. **28**(1), 103–121 (2001)
Turner A, Penn A.: Making Isovists Syntactic: Isovist Integration Analysis. Proceedings 2nd International Symposium on Space Syntax, Brasilia (1999)
Volchenkov, D., Blanchard, P.: Random walks along the streets and canals in compact cities: spectral analysis, dynamical modularity, information, and statistical mechanics. Phys. Rev. E. **75**(2), 6104–6118 (2007)
Weisman, J.: Evaluating architectural legibility: way-finding in the built environment. Environ. Behav. **13**(2), 189–204 (1981)
Wiener, J.M., Franz, G.: Isovists as a means to predict spatial experience and behavior. In: Spati (ed.) Lecture Notes in Artificial Intelligence Number 3343, pp. 42–57. Springer, Berlin (2005)
Wilson, G.: Complex Spatial Systems: the Modelling Foundations of Urban and Regional Analysis. Prentice Hall, Upper Saddle River (2000)
Zachariadis, V.: An agent-based approach to the simulation of pedestrian movement and factors that control it. CUPUM. **2005**, 1–16 (2005)

Hidden Geometry of Urban Landscapes for Smart City Planners

Dimitri Volchenkov

Abstract Urbanization has been the dominant demographic trend in the entire world, during the last half century. Rural to urban migration, international migration, and the reclassification or expansion of existing city boundaries have been among the major reasons for increasing urban population. The essentially fast growth of cities in the last decades urgently calls for a profound insight into the common principles stirring the structure of urban developments all over the world. In the present chapter, we discuss the graph representations of urban spatial structures and suggested a computationally simple technique that can be used in order to spot the relatively isolated locations and neighborhoods, to detect urban sprawl, and to illuminate the hidden community structures in complex urban textures. The approach may be implemented for the detailed expertise of any urban pattern and the associated transport networks that may include many transportation modes.

1 Introduction

A belief in the influence of the built environment on humans was common in architectural and urban thinking for centuries. Cities generate more interactions with more people than rural areas because they are central places of trade that benefit those who live there. People moved to cities because they intuitively perceived the advantages of urban life. City residence brought freedom from customary rural obligations to lord, community, or state and converted a compact space pattern into a pattern of relationships by constraining mutual proximity between people. Spatial organization of a place has an extremely important effect on the way people move through spaces and meet other people by chance (Hillier and Hanson 1984). Compact neighborhoods can foster casual social interactions among neighbors, while creating barriers to interaction with people outside a neighborhood. Spatial

D. Volchenkov (✉)
Mathematics & Statistics, Texas Tech University, Lubbock, TX 79409-1042, USA

Artificial Intelligence Key Laboratory, Sichuan University of Science and Engineering, Sichuan, China
e-mail: dr.volchenkov@gmail.com

S.T. Rassia, P.M. Pardalos (eds.), *Smart City Networks*, Springer Optimization and Its Applications 125, DOI 10.1007/978-3-319-61313-0_12

configuration promotes peoples encounters as well as making it possible for them to avoid each other, shaping social patterns (Ortega-Andeane et al. 2005).

The phenomenon of clustering of minorities, especially that of newly arrived immigrants, is well documented since the work of Wirth (1928) (the reference appears in Vaughan 2005). Clustering is considered to be beneficial for mutual support and for the sustenance of cultural and religious activities. At the same time, clustering and the subsequent physical segregation of minority groups belong to the causes of their economic marginalization. The study of London's change over 100 years performed by Vaughan (2005) has indicated that the creation of poverty areas is a spatial process: by looking at the distribution of poverty at the street, it is possible to find a relationship between spatial segregation and poverty. The patterns of mortality in London studied over the past century by Orford et al. (2002) show that the areas of persistence of poverty cannot be explained other than by an underlying spatial effect.

Urban planning is recognized to play a crucial position in the development of sustainable cities. The essentially fast growth of cities in the last decades urgently calls for a profound insight into the common principles stirring the structure of urban development all over the world.

Sociologists think that isolation worsens an area's economic prospects by reducing opportunities for commerce and engenders a sense of isolation in inhabitants, both of which can fuel poverty and crime. Urban planners and governments have often failed to take such isolation into account when shaping the city landscape, not least because isolation can sometimes be difficult to quantify in the complex fabric of a major city. The source of such a difficulty is profound: while humans live and act in Euclidean space which they percept visually and which is present in them as a mental form, a complex network of interconnected spaces of movements that constitutes a spatial urban pattern does not possess the structure of a Euclidean space (Blanchard and Volchenkov 2009). In Blanchard and Volchenkov (2009) we spoke of fishes: they know nothing either of what the sea, or a lake, or a river might really be and only know the fluid in which they live as if it were air around them. While in a complex built environment, humans have no sensation of it, but need time to construct its "affine representation" so they can understand and store it in their spatial memory. Therefore, human behaviors in complex environments result from a long learning process and the planning of movements within them. In Blanchard and Volchenkov (2009), we suggested that random walks can help us to find such an "affine representation" of the built environment, giving us a leap outside our Euclidean "aquatic surface" and opening up and granting us the sensation of a new space.

While traveling in the city, our primary interest is often in finding the best route from one place to another. Since the wayfinding process is a purposive, directed, and motivated activity (Golledge 1999), the shortest route is not necessarily the best one. If an origin and a destination are not directly connected by a continuous path, wayfinding may include search and exploration actions for which it may be crucial to recognize juxtaposed and distant landmarks, to determine turn angles and directions of movement, and eventually to embed the route in some large reference frame. It is well known that the conceptual representations of space in humans do not bear a one-to-one correspondence with actual physical space. The process of integration of the local affine models of individual places into the entire cognitive map of the urban area network is very complicated and falls largely within the domain of cognitive science and psychology, but nevertheless the specification of what may be recovered from spatial memory can be considered as a problem of mathematics – "the limits of human perception coincide with mathematically plausible solutions" (Pollick 1997). Supposing the inherent mobility of humans and alikeness of their spatial perception aptitudes, one might argue that nearly all people experiencing the city would agree in their judgments on the total number of individual locations in that, in identification of the borders of these locations, and their interconnections. In other words, we assume that spatial experience in humans intervening in the city may be organized in the form of a universally acceptable network.

Well known and frequently traveled path segments provide linear anchors for certain city districts and neighborhoods that help to organize collections of spatial models for the individual locations into a configuration representing the mental image of the entire city. In our study, we assume that the frequently traveled routes are a function of the given layout of streets and squares in the city. It is intuitively clear that if the spatial configuration of the city were represented by a regular graph, where each location represented by a vertex has the same number of neighbors, in the absence of other local landmarks, all paths would be equally probably followed by travelers. No linear anchors are possible in such an urban pattern which could stimulate spatial apprehension. However, if the spatial graph of the city is far from being regular, then a configuration disparity of different places in the city would result in that some of them may be visited by travelers more often than others.

In the following sections of this work, we study the problem of isolation in cities with the use of random walks that provide us with an effective tool for the detailed structural analysis of connected undirected graphs exposing their symmetries (Blanchard and Volchenkov 2009).

2 Spatial Graphs of Urban Environments

In traditional urban researches, the dynamics of an urban pattern come from the landmasses, the physical aggregates of buildings delivering place for people, and their activities. The relationships between certain components of the urban texture are often measured along streets and routes considered as edges of a planar graph, while the traffic end points and street junctions are treated as nodes. Such a primary graph representation of urban networks is grounded on relations between junctions through the segments of streets. The usual city map based on Euclidean geometry can be considered as an example of primary city graphs.

In space syntax theory (see Hillier and Hanson 1984; Hillier 1999), built environments are treated as systems of spaces of vision subjected to a configuration analysis. Being irrelevant to the physical distances, spatial graphs representing the urban environments are removed from the physical space. It has been demonstrated in multiple experiments that spatial perception shapes people understanding of how a place is organized and eventually determines the pattern of local movement (Hillier 1999). The aim of the space syntax study is to estimate the relative proximity between different locations and to associate these distances to the densities of human activity along the links connecting them (Hansen 1959; Wilson 1970; Batty 2004). The surprising accuracy of predictions of human behavior in cities based on the purely topological analysis of different urban street layouts within the space syntax approach attracts meticulous attention (Penn 2001).

The representation of urban spatial networks by connected graphs can be based on a number of different principles. In Jiang and Claramunt (2004), while identifying a street over a plurality of routes on a city map, the named-street approach has been used, in which two different arcs of the primary city network were assigned to the same identification number (ID) provided they share the same street name. In the present chapter, we take a "named-street"-oriented point of view on the decomposition of urban spatial networks into spatial graphs following our previous works (Volchenkov and Blanchard 2007, 2008). Being interested in the statistics of random walks defined on spatial networks of urban patterns, we assign an individual street ID code to each continuous segment of a street. The spatial graph of an urban environment is then constructed by mapping all continuous segments of streets (spaces of motion) into nodes and all intersections between continuous segments of streets into edges.

Although graphs are usually shown diagrammatically, they can also be represented as matrices. The major advantage of a matrix representation is that the analysis of the graph structure can be performed using well-known operations on matrices. For each graph, there exists a unique adjacency matrix (up to permuting rows and columns) which is not the adjacency matrix of any other graph. If we assume that the spatial graph of the city consisting of $i = 1, \ldots, N$ spaces of motion is simple (i.e., it contains neither loops nor multiple edges), the adjacency matrix is a $\{0, 1\}$-matrix with zeros on its diagonal:

$$A_{ij} = \begin{cases} 1, \; i \sim j, & i \neq j, \\ 0, \; \text{otherwise}, \end{cases} \tag{1}$$

where $i \sim j$ means that the space i is directly connected to the space j.

3 Graphs as Discrete Time Dynamical Systems

A finite connected undirected graph can be seen as a *discrete time dynamical system* possessing a finite number of states (nodes) (Prisner 1995). The behavior of such a dynamical system can be studied by means of a transfer operator which describes the time evolution of distributions in state space. The transfer operator can be represented by a stochastic matrix determining a discrete time random walk on the graph in which a walker picks at each node between the various available edges with equal probability. An obvious benefit of the approach based on random walks to graph theory is that the relations between individual nodes and subgraphs acquire a precise quantitative probabilistic description that enables us to attack applied problems which could not even be started otherwise. Given a finite connected undirected graph $G(V, E)$ where V represents the set on graph vertices and E is the set of graph edges, we consider a transformation $\mathcal{S} : V \to V$ mapping any subset of nodes $U \subset V$ into the set of their direct neighbors, $\mathcal{S}(U) = \{w \in V | v \in U, v \sim w\}$. We denote the result of $t \geq 1$ consequent applications of $\mathcal{S}$ to $U \subset V$ by $\mathcal{S}_t(U)$. The iteration of the map $\mathcal{S}$ leads to a study of possible paths in the graph G beginning at some $v \in V$. However, we are rather interested in discussing the time evolution of smooth functions under iteration, than the individual trajectories $\mathcal{S}_t(v)$. Given a discrete *density function* $f(v) \geq 0$, $v \in V$, defined on a undirected connected graph $G(V, E)$ such that $\sum_{v \in V} f(v) = 1$, the dynamics of the map $\mathcal{S}(U)$ is described by the norm-preserving transformation

$$\sum_{v \in U} f(v)\, \mathbf{T}^t = \sum_{\mathcal{S}_t^{-1}(U)} f(v), \tag{2}$$

where $\mathbf{T}^t$ is the *Ruelle–Perron–Frobenius transfer operator* corresponding to the transformation $\mathcal{S}_t$. The uniqueness of the Ruelle–Perron–Frobenius operator for a given transformation $\mathcal{S}_t$ is a consequence of the Radon–Nikodym theorem extending the concept of probability densities to probability measures defined over arbitrary sets (Shilov and Gurevich 1978). It was shown by Mackey (1991) that the relation (2) is satisfied by a homogeneous Markov chain $\{v_t\}_{t \in \mathbb{N}}$ determining a random walk of the nearest neighbor type defined on the connected undirected graph $G(V, E)$ by the transition matrix

$$\begin{aligned} T_{ij} &= \Pr\left[v_{t+1} = j |\, v_t = i\right] > 0 \Leftrightarrow\ i \sim j, \\ &= \mathbf{D}^{-1}\mathbf{A}, \quad \mathbf{D} = \operatorname{diag}(\deg(1), \ldots \deg(N)), \quad \deg(i) \equiv \textstyle\sum_{j=1}^{N} A_{ij}. \end{aligned} \tag{3}$$

where $\mathbf{A}$ is the adjacency matrix of the graph, so that the probability of transition from i to j in $t > 0$ steps is equal to $p_{ij}^{(t)} = (\mathbf{T}^t)_{ij}$. The discrete time random walks on graphs have been studied in details in Lovász (1993), Lovász and Winkler (1995) and by many other authors. For a random walk defined on a connected undirected graph, the Perron–Frobenius theorem asserts the existence of a unique strictly positive probability vector $\boldsymbol{\pi} = (\pi_1, \ldots, \pi_N)$ (the left eigenvector of the transition matrix $\mathbf{T}$ belonging to the maximal eigenvalue $\mu = 1$) such that $\boldsymbol{\pi}\mathbf{T} = 1 \cdot \boldsymbol{\pi}$. For the nearest neighbor random walks defined on an undirected graph, the *stationary distribution* is given by

$$\pi_i = \frac{\deg(i)}{2|E|}, \quad \sum_{i \in V} \pi_i = 1 \tag{4}$$

where deg(i) is the number of immediate neighbors of the node i, and $|E|$ is the total number of edges in the graph. The vector $\boldsymbol{\pi}$ satisfies the condition of detailed balance, $\pi_i T_{ij} = \pi_j T_{ji}$, from which it follows that a random walk defined on an undirected graph is time reversible: it is also a random walk if considered backward, and it is not possible to determine, given the walker at a number of nodes in time after running the walk, which state came first and which state arrived later. The stationary distribution (4) of random walks defined on a connected undirected graph $G(V,E)$ determines a unique measure on V, $D = \sum_{j\in V} \deg(j)\delta_j$ (δ_j being the Dirac measure at j and deg(j) is the degree of the node j, the number of immediate neighbors of j in the graph G) with respect to which the transition operator (3) becomes self-adjoint and is represented by a symmetric transition matrix

$$\widehat{T}_{ij} = \left(\mathbf{D}^{1/2}\,\mathbf{T}\,\mathbf{D}^{-1/2}\right)_{ij} = \frac{A_{ij}}{\sqrt{\deg(i)\deg(j)}}, \qquad \mathbf{D} = \operatorname{diag}\left(\deg(1), \deg(2), \dots, \deg(N)\right). \tag{5}$$

Diagonalizing the symmetric matrix (5), we obtain $\widehat{\mathbf{T}} = \Psi\mathbf{M}\Psi^{\top}$, where Ψ is an orthonormal matrix, $\Psi^{\top} = \Psi^{-1}$, and $\mathbf{M}$ is a diagonal matrix with entries $1 = \mu_1 > \mu_2 \geq \ldots \geq \mu_N > -1$ (here, we exclude bipartite graphs, for which $\mu_N = -1$). The rows $\boldsymbol{\psi}_k = \{\psi_{k,1}, \dots, \psi_{k,N}\}$ of the orthonormal matrix Ψ forms an orthonormal basis in Hilbert space $\mathscr{H}(V)$, $\boldsymbol{\psi}_k : V \to S_1^{N-1}$, $k = 1, \dots N$, where S_1^{N-1} is the $N-1$-dimensional unit sphere. We consider the eigenvectors $\boldsymbol{\psi}_k$ ordered in accordance to the eigenvalues they belong to. For eigenvalues of algebraic multiplicity $\alpha > 1$, a number of linearly independent orthonormal ordered eigenvectors can be chosen to span the associated eigenspace. The first eigenvector $\boldsymbol{\psi}_1$ belonging to the largest eigenvalue $\mu_1 = 1$ (which is simple) is the Perron–Frobenius eigenvector that determines the stationary distribution of random walks over the graph nodes, $\psi_{1,i}^2 = \pi_i$, $i = 1, \dots, N$. The squared Euclidean norm of the vector in the orthogonal complement of $\boldsymbol{\psi}_1$, $\sum_{s=2}^{N} \psi_{s,i}^2 = 1 - \pi_i > 0$ expresses the probability that a random walker is not in i.

4 Affine Probabilistic Geometry of Eigenvector Embedding and Probabilistic Euclidean Distance on Graphs

Discovering of important nodes and quantifying differences between them in a graph is not easy, since the graph does not possess a priori the structure of a Euclidean space. We use the algebraic properties of the self-adjoint operators in order to define an Euclidean metric on any finite connected undirected graph. Geometric objects, such as points, lines, or planes, can be given a representation as elements in a projective space based on homogeneous coordinates. Given an orthonormal basis $\left\{\boldsymbol{\psi}_k : V \to S_1^{N-1}\right\}_{k=1}^{N}$ in $\mathbb{R}^N$, any vector in Euclidean space can be expanded into $\mathbf{v} = \sum_{k=1}^{N} \langle \mathbf{v}|\psi_k\rangle \langle \psi_k|$. Provided $\{\boldsymbol{\psi}_k\}_{k=1}^{N}$ are the eigenvectors of the symmetric matrix of the operator $\widehat{\mathbf{T}}$, we can define the new basis vectors, $\psi' \equiv \left\{1, \frac{\psi_{2,2}}{\psi_{1,2}}, \ldots, \frac{\psi_{N,N}}{\psi_{1,N}}\right\}$, since we have $\psi_{1,i} \equiv \sqrt{\pi_i} > 0$ for any $i \in V$. The new basis vectors $\psi'_k := \left(1, \frac{\psi_{k,2}}{\psi_{1,2}}, \ldots, \frac{\psi_{k,N}}{\psi_{1,N}}\right)$, $k = 2, \ldots, N$, span the projective space $P\mathbb{R}_\pi^{(N-1)}$, so that the vector $\mathbf{v}$ can be expanded into

$$\mathbf{v}\boldsymbol{\pi}^{-1/2} = \sum_{k=2}^{N} \langle v|\psi'_k\rangle\langle\psi'_k| . \tag{6}$$

It is easy to see that the transformation (6) defines a stereographic projection onto $P\mathbb{R}_\pi^{(N-1)}$ such that all vectors in $\mathbb{R}^N(V)$ collinear to the vector $|\ \psi_1\rangle$, corresponding to the stationary distribution of random walks, are projected onto a common image point. If the graph $G(V, E)$ has some isolated nodes $\iota \in V$, for which $\pi_\iota = 0$, they play the role of the plane at infinity with respect to (6), away from which we can use the basis Ψ' as an ordinary Cartesian system. The transition to the homogeneous coordinates transforms vectors of $\mathbb{R}^N$ into vectors on the $(N-1)$-dimensional hypersurface $\left\{\psi_{1,i} = \sqrt{\pi_i}\right\}$, the orthogonal complement to the vector of stationary distribution $\boldsymbol{\pi}$.

The Green function (a pseudo-inverse) of the normalized Laplace operator $\widehat{\mathbf{L}} = \mathbf{1} - \widehat{\mathbf{T}}$ describing the diffusion of random walkers over the undirected graph is given in the homogeneous coordinates by $\widehat{\mathbf{L}}^{\natural} = \sum_{k=2}^{N} \lambda_k^{-1} |\psi'_k\rangle\langle\psi'_k|$, where $\lambda_k := 1 - \mu_k$ are the eigenvalues of the self-adjoint operator $\widehat{\mathbf{L}}$. In order to obtain a Euclidean metric on the graph $G(V, E)$, one needs to introduce distances between points (nodes of the graph) and the angles between vectors pointing at them that can be done by determining the dot product $(\boldsymbol{\xi}, \boldsymbol{\zeta})_T$ between any two vectors $\boldsymbol{\xi}, \boldsymbol{\zeta} \in P\mathbb{R}_\pi^{(N-1)}$ by $(\boldsymbol{\xi}, \boldsymbol{\zeta})_T := \left(\boldsymbol{\xi}, \mathbf{L}^{\natural}\boldsymbol{\zeta}\right)$, where $(\cdot, \cdot)$ is the standard scalar product in $P\mathbb{R}_\pi^{(N-1)}$. The dot product is a symmetric real-valued scalar function that allows us to define the (squared) norm of a vector $\boldsymbol{\xi} \in P\mathbb{R}_\pi^{(N-1)}$ by $\|\ \boldsymbol{\xi}\ \|_T^2 := \left(\boldsymbol{\xi}, \mathbf{L}^{\natural}\boldsymbol{\xi}\right)$. The squared norm

$$\|\ \mathbf{e}_i\ \|_T^2 = \frac{1}{\pi_i} \sum_{s=2}^{N} \frac{\psi_{s,i}^2}{\lambda_s} \tag{7}$$

of the canonical basis vector $\mathbf{e}_i = \{0, \ldots 1_i, \ldots 0\}$ representing the node $i \in V$ is nothing else, but the spectral representation of the *first passage time* to the node $i \in V$, the expected number of steps required to reach the node $i \in V$ for the first time starting from a node randomly chosen among all nodes of the graph accordingly to the stationary distribution π. The first passage time, $\|\mathbf{e}_i\|_T^2$, can be directly used in order to characterize the level of accessibility of the node i (Blanchard and Volchenkov 2011). The Euclidean distance between two vectors, $\mathbf{e}_i$ and $\mathbf{e}_j$, given by $\left\|\mathbf{e}_i - \mathbf{e}_j\right\|_T^2 = \|\ \mathbf{e}_i\ \|_T^2 + \left\|\ \mathbf{e}_j\ \right\|_T^2 - 2\left(\mathbf{e}_i, \mathbf{e}_j\right)_T$ is nothing else, but the *commute time*, the expected number of steps required for a random walker starting at $i \in V$ to visit $j \in V$ and then to return back to i (Blanchard and Volchenkov 2011).

5 First-Passage Times to Ghettos

The phenomenon of clustering of minorities, especially that of newly arrived immigrants, is well documented (Wirth 1928). Clustering is considered to be beneficial for mutual support and for the sustenance of cultural and religious activities. At the same time, clustering and the subsequent physical segregation of minority groups would cause their economic marginalization. The spatial analysis of the immigrant quarters (Vaughan 2005) and the study of London's changes over 100 years (Vaughan et al. 2005) show that they were significantly more segregated from the neighboring areas; in particular, the number of street turning away from the quarters to the city centers were found to be less than in the other inner city areas being usually socially barricaded by railways, canals, and industries. It has been suggested (Hillier 2004) that the space structure and its impact on movement are critical for the link between the built environment and its social functioning. Spatial structures creating a local situation in which there is no relation between movements inside the spatial pattern and outside it and the lack of natural space occupancy become associated with the social misuse of the structurally abandoned spaces. We have analyzed the first-passage times to individual canals in the spatial graph of the canal network in Venice. The distribution of numbers of canals over the range of the first-passage time values is represented by a histogram shown in Fig. 1(left). The height of each bar in the histogram is proportional to the number of canals in the canal network of Venice for which the first-passage times fall into the disjoint intervals (known as bins). Not surprisingly, the Grand Canal, the giant Giudecca Canal, and the Venetian Lagoon are the most connected. In contrast, the Venetian Ghetto (see Fig. 1(right)) – jumped out as by far the most isolated, despite being apparently well connected to the rest of the city – on average, it took 300 random steps to reach, far more than the average of 100 steps for other places in Venice.

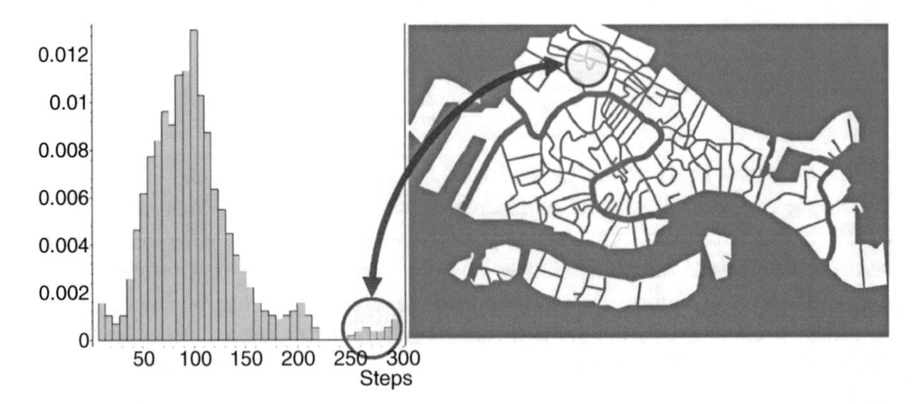

Fig. 1 The Venetian Ghetto jumped out as by far the most isolated, despite being apparently well connected to the rest of the city

6 Why Is Manhattan so Expensive?

The notion of isolation acquires a statistical interpretation by means of random walks. The recurrence time of random walks to a node in an undirected graph depends upon the connectivity of the node, the number of its nearest neighbors – the local structural property of the node in the graph. On the contrary, the first-passage time to the node, the expected number of steps required for a random walker to reach the node from any other node chosen over the graph accordingly the stationary distribution of walks, represents the global structural property of the node in the graph, as all possible paths to it by a self-avoiding random walk (with no self-loops) are taken into account although some paths are rendered more probable than others. While the relatively low recurrence times might be typical for highly connected nodes disregarding their role for the entire graph structure, the relatively low first-passage times indicate the importance of nodes aggregating many a path of all lengths for structural integrity of the entire graph even if their connectivity is relatively low. For example, a bridge connecting the city districts situated on the opposite banks of a river is vital for the entire urban transportation system despite its limited connectivity to the immediate city neighborhoods (Blanchard and Volchenkov 2009). The spaces of motion in the spatial graph characterized by the shortest first-passage times are easy to reach by whatever origin-destination route while very many random steps would be required in order to get into a statistically isolated site. Being a global characteristic of a node in the graph, the first-passage time assigns absolute scores to all nodes based on the probability of paths they provide for random walkers. The first-passage time can therefore be considered as a natural statistical centrality measure of the node within the graph (Blanchard and Volchenkov 2009).

A visual pattern displayed on Fig. 2 represents the pattern of structural isolation (quantified by the first-passage times) in Manhattan (darker color corresponds to longer first-passage times). It is interesting to note that the spatial distribution of isolation in the urban pattern of Manhattan (Fig. 2) shows a qualitative agreement with the map of the tax assessment value of the land in Manhattan reported by B. Rankin (2006) in the framework of the RADICAL CARTOGRAPHY project. The first-passage times enable us to classify all places in the spatial graph of Manhattan into four groups accordingly to the first-passage times to them (Blanchard and Volchenkov 2009). The first group of locations is characterized by the minimal first-passage times; they are probably reached for the first time from any other place of the urban pattern in just 10–100 random navigational steps (the heart of the city); see Figs. 3(1) and (2). These locations are identified as belonging to the downtown of Manhattan (at the south and southwest tips of the island) – the Financial District and Midtown Manhattan. It is interesting to note that these neighborhoods are roughly coextensive with the boundaries of the ancient New Amsterdam settlement founded in the late seventeenth century. Both districts comprise the offices and headquarters of many of the city's major financial institutions such as the New York Stock Exchange and the American Stock Exchange (in the Financial District). Federal Hall

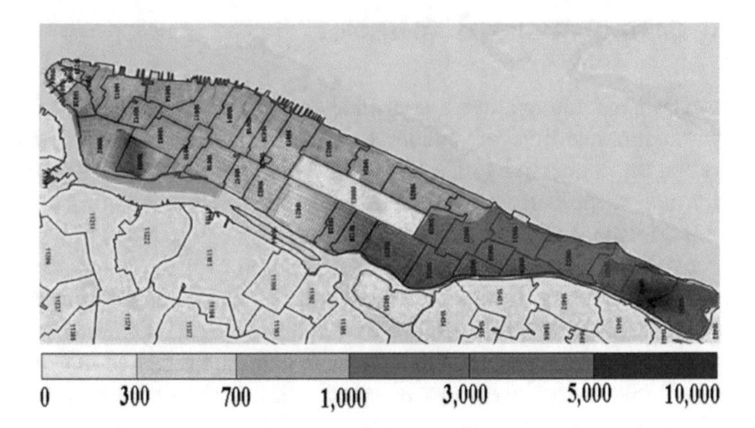

Fig. 2 Isolation map of Manhattan. Isolation is measured by first-passage times to the places in Manhattan calculated from the randomly chosen place with respect to the stationary distribution of random walks. Darker color corresponds to longer first-passage times

National Memorial is also encompassed in this area that had been anchored by the World Trade Center until the September 11, 2001 terrorist attacks. We might conclude that the group of locations characterized by the best structural accessibility is the heart of the public process in the city.

The neighborhoods from the second group (the city core) comprise the locations that can be reached for the first time in several hundreds to roughly a thousand random navigational steps from any other place of the urban pattern (Fig. 3(3) and (4)). SoHo (to the south of Houston Street), Greenwich Village, Chelsea (Hell's Kitchen), the Lower East Side, and the East Village are among them – they are commercial in nature and known for upscale shopping and the "Bohemian" lifestyle of their dwellers contributing into New York's art industry and nightlife.

The relatively isolated neighborhoods such as Bowery (Fig. 3(5)), some segments in Hamilton Heights and Hudson Heights, Manhattanville (bordered on the south by Morningside Heights), TriBeCa (Triangle Below Canal), and some others can be associated to the third structural category as being reached for the first time from 1,000 to 3,000 random steps starting from a randomly chosen place in the spatial graph of Manhattan. Interestingly, that many locations belonging to the third structural group comprise the diverse and eclectic mix of different social and religious groups. Many famous houses of worship had been established there during the late nineteenth century – St. Mary's Protestant Episcopal Church, Church of the Annunciation, St. Joseph's Roman Catholic Church, and Old Broadway Synagogue in Manhattanville are among them. The neighborhood of Bowery in the southern portion of Manhattan had been most often associated with the poor and the homeless. From the early twentieth century, Bowery became the center of the so-called "b'hoy" subculture of working-class young men frequenting the cruder nightlife. Petty crime and prostitution followed in their wake, and most respectable businesses, the middle-class, and entertainment had fled the area. Nowadays, the

Fig. 3 The first passage times in the borough of Manhattan, NYC: (*1*) the Federal Hall National Memorial ∼10 steps; (*2*) the Times square ∼100 steps; (*3*) the SoHo neighborhood, in Lower Manhattan ∼500 steps; (*4*) the East Village neighborhood, lying east of Greenwich Village, south of Gramercy and Stuyvesant Town ∼1,000 steps; (*5*) the Bowery neighborhood, in the southern portion of the New York City borough of Manhattan ∼5,000 steps; (*6*) the East Harlem (Spanish Harlem, El Barrio), a section of Harlem located in the northeastern extremity of the borough of Manhattan ∼10,000 steps

dramatic decline has lowered crime rates in the district to a level not seen since the early 1960s and continue to fall. Although zero-tolerance policy targeting petty criminals is being held up as a major reason for the crime combat success, no clear explanation for the crime rate fall has been found.

The last structural category comprises the most isolated segments in the city mainly allocated in the Spanish and East Harlem. They are characterized by the longest first-passage times from 5,000 to 10,000 random steps starting from a randomly chosen place in the spatial graph of Manhattan (Fig. 3(6)). Structural isolation is fostered by the unfavorable confluence of many factors such as the close proximity to Central Park (an area of 340 hectares removed from the otherwise regular street grid), the Harlem River separating the Harlem and the Bronx, and the remoteness from the main bridges (the Triborough Bridge, the Willis Avenue Bridge, and the Queensboro Bridge) that connect the boroughs of Manhattan to the urban arrays in Long Island City and Astoria. Many social problems associated with poverty from crime to drug addiction have plagued the area for some time. The haphazard change of the racial composition of the neighborhood occurred at the beginning of the twentieth century together with the lack of adequate urban infrastructure and services fomenting racial violence in deprived communities and made the neighborhood unsafe – Harlem became a slum. The neighborhood had suffered with unemployment, poverty, and crime for quite long time and even now, despite the sweeping economic prosperity and redevelopment of many sections in the district, the core of Harlem remains poor.

Recently, we have discussed in Blanchard and Volchenkov (2011) that distributions of various social variables (such as the mean household income and prison expenditures in different zip code areas) may demonstrate the striking spatial patterns which can be analyzed by means of random walks. In the present work, we analyze the spatial distribution of the tax assessment rate (TAR) in Manhattan.

The assessment tax relies upon a special enhancement made up of the land or site value and differs from the market value estimating a relative wealth of the place within the city commonly referred to as the "unearned" increment of land use, Bolton (1922). The rate of appreciation in value of land is affected by a variety of conditions, for example, it may depend upon other property in the same locality, will be due to a legitimate demand for a site, and for occupancy and height of a building upon it.

The current tax assessment system enacted in 1981 in the city of New York classifies all real estate parcels into four classes subjected to the different tax rates set by the legislature: (i) primarily residential condominiums; (ii) other residential property; (iii) real estate of utility corporations and special franchise properties; (iv) all other properties, such as stores, warehouses, hotels, etc. However, the scarcity of physical space in the compact urban pattern on the island of Manhattan will naturally set some increase of value on all desirably located land as being a restricted commodity. Furthermore, regulatory constraints on housing supply exerted on housing prices by the state and the city in the form of "zoning taxes" are responsible for converting the property tax system in a complicated mess of interlocking influences and for much of the high cost of housing in Manhattan (Glaeser and Gyourko 2003).

Being intrigued with the likeness of the tax assessment map and the map of isolation in Manhattan, we have mapped the TAR figures publicly available through the Office of the Surveyor at the Manhattan Business Center onto the data on first-passage times to the corresponding places. The resulting plot is shown in Fig. 4, in the logarithmic scale. The data presented in Fig. 4 positively relates the geographic accessibility of places in Manhattan with their "unearned increments" estimated by means of the increasing burden of taxation. The inverse linear pattern dominating the data is best fitted by the simple hyperbolic relation between the tax assessment rate (TAR) and the value of first-passage time (FPT), $\mathrm{TAR} \approx 1.2 \cdot 10^5/\mathrm{FPT}$.

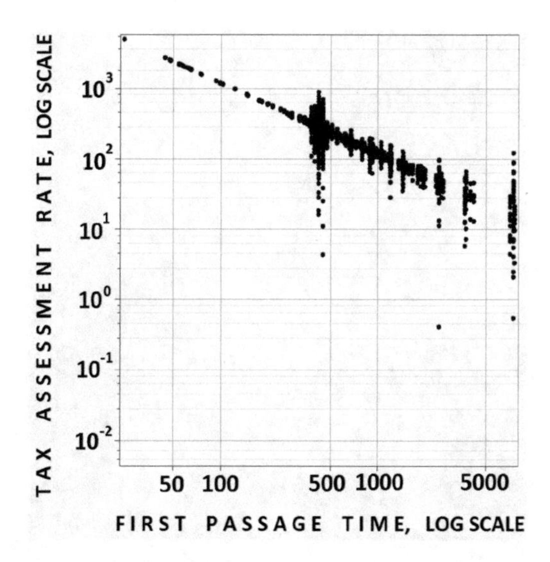

Fig. 4 Tax assessment rate (TAR) of places in Manhattan (the vertical axes is scaled in dollars per square foot) is shown in the logarithmic scale vs. the first–passage times (FPT) to them (the horizontal axes)

7 Mosque and Church in Dialog

Churches are buildings used as religious places, in the Christian tradition. In addition to being a place of worship, the churches in Western Europe were utilized by the community in other ways, e.g., they could serve as a meeting place for guilds. Typically, their location were at a focus of a neighborhood, or a settlement (Fig. 5).

Nowadays, because of the intensive movement of people between countries, the new national unities out of cultural and religious diversity have appeared. The USA possessing rich tradition of immigrants has demonstrated the ability of an increasingly multicultural society to unite different religious, ethnic, and linguistic groups into the fabric of the country, and many European countries follow that way (Portes and Rumbaut 2006).

Religious beliefs and institutions have played and continue to play a crucial role in new immigrant communities. Religious congregations often provide ethnic, cultural, and linguistic reinforcements and often help newcomers to integrate by offering a connection to social groups that mediate between the individual and the new society, so that immigrants often become even more religious once in the new country of residence (Kimon 2001).

It is not a surprise that the buildings belonging to religious congregations of newly arrived immigrants are usually located not at the centers of cities in the host country – the changes in function results in a change of location. In the previous section, we have discussed that religious organizations of immigrants in the urban pattern of Manhattan have been usually founded in the relatively isolated locations, apart from the city core, like those in Manhattanville. We have seen that the typical first-passage times to the "religious" places of immigrant communities in Manhattan

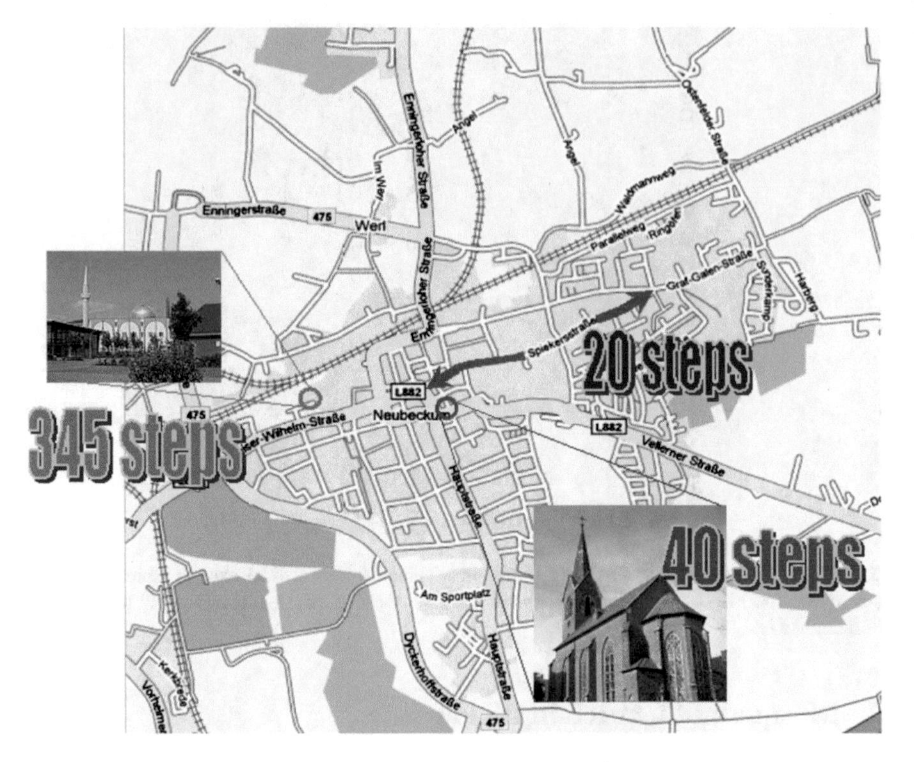

Fig. 5 Neubeckum (Westphalia): the church and the mosque in dialog

scale from 1,000 to 3,000 random steps (Blanchard and Volchenkov 2009). It is interesting to check this observation also for the religious congregation buildings of recent immigrants in Western Europe.

Let us discuss briefly an example concerning a town in the northern part of North Rhine-Westphalia, Germany. Despite the mosque and the church are located in close geographic proximity in the city of Neubeckum, their locations are dramatically different with respect to the entire city structure. The analysis of the spatial graph of the city of Neubeckum by random walks shows that while the church is situated in a place belonging to the city core, and just 40 random steps are required in order to reach it for the first time from any arbitrary chosen place, a random walker needs 345 random steps to arrive at the mosque. The commute time, the expected number of steps a random walker needs to reach the mosque from the church and then to return back, equals 405 steps.

Spiekersstrasse, the street which is parallel to the railway, now is the best accessible place of motion in Neubeckum playing the role of its structural "center of mass"; it can be achieved from any other location in the city in just 20 random steps. The relation between the extent of structural isolation and the specified reference levels can be measured in a logarithmic scale by using as unit of decibel (dB) (Blanchard and Volchenkov 2009). When referring to estimates of isolation by means of first-passage times (FPT), a ratio between two levels inherent to the different locations A and B can be expressed in decibels by evaluating

$$\mathrm{I}_{AB} = 10 \log_{10} \left(\frac{\mathrm{FPT}(A)}{\mathrm{FPT}(B)} \right),$$

where FPT(A) and FPT(B) are the first-passage times to A and B, respectively. If we estimate relative isolation of other places of motion with respect to Spiekersstrasse by comparing their first-passage times in the logarithmic scale, then the location of the church is evaluated by $I_{\mathrm{Church}} \approx 3\,\mathrm{dB}$ of isolation and $I_{\mathrm{Mosque}} \approx 12\,\mathrm{dB}$ for the mosque.

Indeed, isolation was by no means the aim of the Muslim community. The mosque in Neubeckum has been erected on a vacant place, where land is relatively cheap. However, structural isolation under certain conditions would potentially have dramatic social consequences. Efforts to develop systematic dialogue and increased cooperation based on a reinforced culture of consultations are viewed as essential to deliver a sustainable community.

8 Which Place Is the Ideal Bielefeld Crime Scene?

Bielefeld is a city in the northeast of North-Rhine Westphalia (Germany), famous as a home to a significant number of internationally operating companies.

"Which place is the ideal Bielefeld crime scene?" This question has been recently addressed by the *Bielefeld–heute* (*"Bielefeld today"*) weekly newspaper to those crime fiction authors who had chosen Bielefeld as a stage for the criminal stories of their novels. Although the above question falls largely within the domain of criminal psychology, it can also be considered as a problem of mathematics – since the limits of human perception coincide with mathematically plausible solutions.

We have analyzed how easy it is to get to various places on the labyrinth in the network of 200 streets located at the city center of Bielefeld aiming to capture a neighborhood's inaccessibility which could expose hidden islands of future deprivation and social misuse in that. For our calculations, we imagined pedestrians wandering randomly along the streets and worked out the average number of random turns at junctions they would take to reach any particular place in Bielefeld from various starting points. Not surprisingly, the *August-Bebel Str.*, *Dorotheenstraße*, and the *Herforder Str.* were the most accessible in the city. In contrast, we found that the certain districts located along the railroad (see the map shown in Fig. 6) jumped out as being by far the most isolated, despite being apparently well connected to the rest of the city.

On average, it took from 1,389 to 1,471 random treads to reach such the godforsaken corners as the parking places on *Am Zwinger* (Fig. 7(1)), the neighborhood

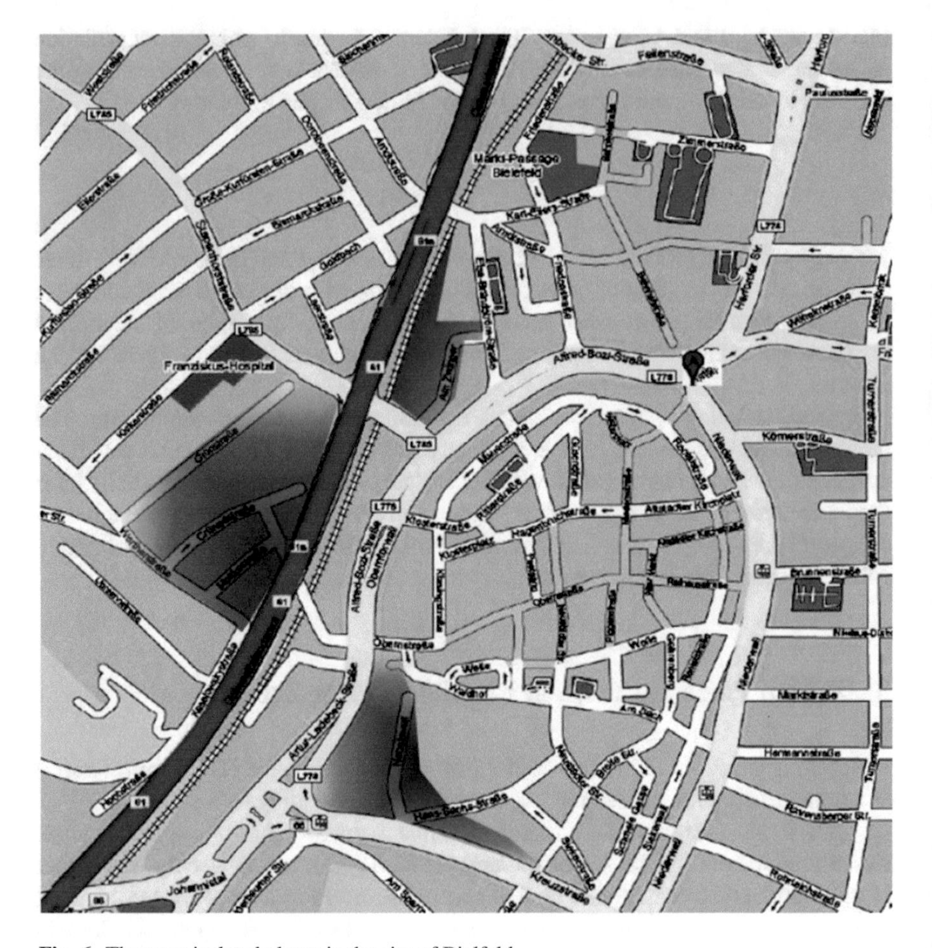

Fig. 6 The most isolated places in the city of Bielfeld

centered by the *Crüwellstraße* (Fig. 7(2)), the waste places close to the Natural History Museum, and the City Art Gallery (Fig. 7(3)) – far more than the average of 450 steps for other places in Bielefeld.

The inhospitable isolation of these places can be estimated numerically; however, people rather percept it intuitively. Although the actual criminal rate in Bielefeld appears to be relatively low, many pulp fiction authors found the city a suitable place for their criminal stories that indeed recalls us the sustained satirical Internet myth of the *Bielefeld Verschwörung* (Bielefeld Conspiracy) going that the city of Bielefeld in the German State of North Rhine-Westphalia does not actually exist. In spite of all efforts to subsidize development and publicity for Bielefeld by the city council trying hard to build a nationwide known public image of the city, the mayor's office still reportedly receives phone calls and e-mails which claim to doubt the existence of the city (Article 2012).

Fig. 7 The ideal Bielefeld crime scene: (*1*) the parking place *Am Zwinger*, (*2*) the intersection of *Crüwellstraße* and *Moltkestraße*, and (*3*) the place beyond the Natural History Museum and the City Art Gallery

Our analysis shows that the city of Bielefeld consists of three structurally different components loosely tied together by just a few principal routes. Being founded in 1214 by Hermann IV, the Count of Ravensberg, the compact city guarded a pass crossing the Teutoburg Forest. In 1847, the new Cologne-Minden Railway had passed through Bielefeld establishing the new urban development apart from the historical core of the major city – the Bahnhofsviertel governed mostly by the linear structure of the railroad. Finally, during the industrial revolution, the modern city quarters had been constructed by the end of the nineteenth century. These city districts built in accordance with different development principles and in different historical epochs are strikingly dissimilar in structure. Walkers in our model were mostly confined in each city domain experiencing difficulty while alternating that. Not surprisingly, most Germans have a vague image of the city in their heads. The threefold structure of Bielefeld would make the city center extremely vulnerable to proliferation of growth problems.

9 Conclusion and Discussion

We assumed that spatial experience in humans intervening in the city may be organized in the form of a universally acceptable network. We also assumed that the frequently traveled routes are nothing else but the "projective invariants" of the given layout of streets and squares in the city – the function of its geometrical configuration, which remains invariant whatever origin–destination route is considered.

Basing on these two assumptions, we have developed a method that allows to capture a neighborhood's inaccessibility. Any finite undirected graph can be interpreted as a discrete time dynamical system with a finite number of states. The temporal evolution of such a dynamical system is described by a "dynamical law" that maps vertices of the graph into other vertices and can be interpreted as the transition operator of random walks. The level of accessibility of nodes and subgraphs of undirected graphs can be estimated precisely in connection with random walks introduced on them. We have applied this method to the structural analysis of different cities.

The main motivation of our work was to get an insight into the structure of human settlements that would improve the overall strategy of investments and planning and avoid the declining of cities as well as reduce many environmental problems. Multiple increases in urban population that had occurred in Europe at the beginning of the twentieth century have been among the decisive factors that changed the world. Urban agglomerations had suffered from the comorbid problems such as widespread poverty, high unemployment, and rapid changes in the racial composition of neighborhoods. Riots and social revolutions have occurred in urban places in many European countries in part in response to deteriorated conditions of urban decay and fostered political regimes affecting immigrants and certain population groups de facto alleviating the burden of the haphazard urbanization by increasing its deadly price.

Urbanization has been the dominant demographic trend in the entire world, during the last half century. Although the intense process of urbanization is a proof of economic dynamism, clogged roads, dirty air, and deteriorating neighborhoods are fueling a backlash against urbanization that nevertheless cannot be stopped. The urban design decisions made today on the base of the US car-centered model, in cities of the developing world where car use is still low, will have an enormous impact on climate changes in the decades ahead. Unsustainable pressure on resources causes the increasing loss of fertile lands through degradation, and the dwindling amount of fresh water and food would trigger conflicts and result in mass migrations. Migrations induce a dislocation and disconnection between the population and their ability to undertake traditional land use (Fisher 2008). Major metropolitan areas and the intensively growing urban agglomerations attract large numbers of immigrants with limited skills. Many of them will end up a burden on the state and perhaps become involved in criminal activity. The poor are urbanizing faster than the population as a whole (Ravallion 2007). Global poverty is in flight becoming a primarily urban phenomenon in the developing world: about 70% of 2 bln new urban settlers in the next 30 years will live in slums, adding to 1 bln already there (Blanchard and Volchenkov 2009). The essential attention should be given to the cities in the developing world where the accumulated urban growth will be duplicated in the next 25 years. The fastest urbanization of poverty occurred in Latin America, where the majority of the poor now live in urban areas.

Faults in urban planning, poverty, redlining, immigration restrictions, and clustering of minorities dispersed over the spatially isolated pockets of streets trigger urban decay, a process by which a city falls into a state of disrepair. The speed and scale of urban growth require urgent global actions to help cities prepare for growth and to avoid them of being the future epicenters of poverty and human suffering.

People of modern Europe prefer to live in single-family houses and commute by automobile to work. In 10 years (1990–2000), low-density expansions of urban areas known as "urban sprawl" consumed more than 8,000 km^2 in Europe, the entire territory of the state of Luxembourg. Residents of sprawling neighborhoods tend to emit more pollution per person and suffer more traffic fatalities. Faults in planning of urban sprawl neighborhoods would force the structural focus of the city out from its historical center and trigger the process of degradation in that.

Together with severe environmental problems generated by the unlimited expansion of the city, the process of urban degradation creates dramatic economic and social implications, with negative effects on the urban economy. It is well known that degraded urban areas are less likely to attract investments, new enterprises, and services, but become attractive for socially underprivileged groups because of a tendency of reduction house prices in the urban core. Smart growth policies that concentrate the future urban development in the center of the city to avoid urban sprawl should be applied.

Our last but not least remark is that sprawling suburbs in the USA saw by far the greatest growth in their poor population and by 2008 had become home to the largest share of the nation's poor. Between 2000 and 2008, sprawls in the US largest metro areas saw their poor population grow by 25% – almost five times faster than primary cities and well ahead of the growth seen in smaller metro areas and non-metropolitan communities. These trends are likely to continue in the wake of the latest downturn, given its toll on the faster pace of growth in suburban unemployment.

A combination of interrelated factors, including urban planning decisions, poverty, the development of freeways and railway lines, suburbanization, redlining, and immigration restrictions, would trigger urban decay, a process by which a city falls into a state of disrepair. We often think that we have much enough time on our hands, but do we? The need could not be more urgent and the time could not be more opportune to act now to sustain our common future.

Acknowledgements The author gratefully acknowledges the financial support by the Cluster of Excellence Cognitive Interaction Technology "CITEC" (EXC 277) at Bielefeld University, which is funded by the German Research Foundation (DFG).

References

Article: Auch Merkel zweifelt an Existenz Bielefelds (German)(Even Merkel doubts the existence of Bielefeld), Die Welt, November 27 (2012)

Batty, M.: A New Theory of Space Syntax, UCL Centre For Advanced Spatial Analysis Publications, CASA Working Paper 75 (2004)

Blanchard, Ph., Volchenkov, D.: Introduction to Random Walks on Graphs and Databases. Springer Series in Synergetics, vol. 10. Springer, Berlin/Heidelberg (2011). http://dx.doi.org/10.1007/978-3-642-19592-1. ISBN:978-3-642-19591-4

Blanchard, Ph., Volchenkov, D.: Mathematical Analysis of Urban Spatial Networks. Springer Series Understanding Complex Systems, 181 pp. Springer, Berlin/Heidelberg (2009). ISBN:978-3-540-87828-5

Bolton, R.P.: Building For Profit. De Vinne Press, New York (1922)

Fisher, M.: Urban ecology. Permaculture Design course handout notes. Available at www.self-willed-land.org.uk (2008)

Glaeser, E.L., Gyourko, J.: Why is manhattan so expensive? Manhattan Institute for Policy Research. Civic Report **39** (2003)

Golledge, R.G.: Wayfinding Behavior: Cognitive Mapping and Other Spatial Processes. John Hopkins University Press, Baltimore (1999). ISBN:0-8018-5993-X

Hansen, W.G.: How accessibility shapes land use. J. Am. Inst. Planners **25**, 73 (1959)

Hillier, B.: Space is the Machine: a Configurational Theory of Architecture. Cambridge University Press (1999). ISBN 0-521-64528-X

Hillier, B.: The common language of space: a way of looking at the social, economic and environmental functioning of cities on a common basis, Bartlett School of Graduate Studies, London (2004)

Hillier, B., Hanson, J.: The Social Logic of Space (1993, reprint, paperback edition ed.). Cambridge University Press, Cambridge (1984)

Jiang, B., Claramunt, C.: Topological analysis of urban street networks. Environ. Plann. B. Plann. Des. **31**, 151 (2004). Pion Ltd.

Kimon, H.: Religion and New Immigrants: a Grantmaking Strategy at The Pew Charitable Trusts. Religion Program, the Pew Charitable Trusts (2001)

Lovász, L.: Random Walks On Graphs: a Survey. Bolyai Society Mathematical Studies 2: Combinatorics, Paul Erdös is Eighty, 1–46. Keszthely (Hungary) (1993)

Lovász, L., Winkler, P.: Mixing of random walks and other diffusions on a graph. Surveys in Combinatorics, Stirling. London Mathematical Society Lecture Note Series, vol. 218, pp. 119–154. Cambridge University Press, Cambridge (1995)

Mackey, M.C.: Time's Arrow: the Origins of Thermodynamic Behavior. Springer, New York (1991)

Orford, S., Dorling, D., Mitchell, R., Shaw, M., Davey-Smith, G.: Life and death of the people of London: a historical GIS of Charles Booth's inquiry. Health and Place **8**(1), 25–35 (2002)

Ortega-Andeane, P., Jiménez-Rosas, E., Mercado-Doménech, S., Estrada-Rodrýguez, C.: Space syntax as a determinant of spatial orientation perception. Int. J. Psychol. **40**(1), 11–18 (2005)

Penn, A.: Space Syntax and Spatial Cognition. Or, why the axial line? In: Peponis, J., Wineman, J., Bafna, S. (eds.) Proceedings of the Space Syntax 3rd International Symposium. Georgia Institute of Technology, Atlanta (2001)

Pollick, F.E.: The perception of motion and structure in structure-from-motion: comparison of affine and Euclidean formulations. Vis. Res. **37**(4), 447–466 (1997)

Portes, A., Rumbaut, R.G.: Immigrant America: a Portrait, 3rd edn. University of California Press, Berkeley (2006)

Prisner, E.: Graph Dynamics. CRC Press, Boca Raton (1995)

Ravallion, M.: Urban poverty. Financ. Dev. **44**(3), 15–17 (2007)

Shilov, G.E., Gurevich, B.L.: Integral, Measure, and Derivative: A Unified Approach (trans: Silverman, R.A. from Russian). Dover Publications, New York (1978)

Vaughan, L.: The relationship between physical segregation and social marginalization in the urban environment. World Architecture **185**, 88–96 (2005)

Vaughan, L., Chatford, D., Sahbaz, O.: Space and Exclusion: The Relationship between physical segregation. Economic marginalization and poverty in the city. Paper presented to Fifth Intern. Space Syntax Symposium, Delft, Holland (2005)

Volchenkov, D., Blanchard, Ph.: Random walks along the streets and channels in compact cities: spectral analysis, dynamical modularity, information, and statistical mechanics. Phys. Rev. E **75**, 026104 (2007)

Volchenkov, D., Blanchard, Ph.: Scaling and universality in city space syntax: between Zipf and Matthew. Phys. A **387**(10), 2353 (2008). doi:10.1016/j.physa.2007.11.049

Wilson, A.G.: Entropy in Urban and Regional Modeling. Pion Press, London (1970)

Wirth, L.: The Ghetto (edition 1988) Studies in Ethnicity. Transaction Publishers, New Brunswick/London (1928)

Printed in the United States
By Bookmasters